Biodiversity
An Introduction

BIODIVERSITY

An Introduction

KEVIN J. GASTON
Royal Society University Research Fellow

AND

JOHN I. SPICER
Lecturer in Zoology

Department of Animal & Plant Sciences
University of Sheffield

**Blackwell
Science**

© 1998 by
Blackwell Science Ltd
Editorial Offices:
Osney Mead, Oxford OX2 0EL
25 John Street, London WC1N 2BL
23 Ainslie Place, Edinburgh EH3 6AJ
350 Main Street, Malden
   MA 02148 5018, USA
54 University Street, Carlton
   Victoria 3053, Australia
10, rue Casimir Delavigne
   75006 Paris France

Other Editorial Offices:
Blackwell Wissenschafts-Verlag GmbH
Kurfürstendamm 57
10707 Berlin, Germany

Blackwell Science KK
MG Kodenmacho Building
7–10 Kodenmacho Nihombashi
Chuo-ku, Tokyo 104, Japan

First published 1998

Set by Excel Typesetters, Hong Kong
Printed and bound in Great Britain
by MPG Books Ltd, Bodmin, Cornwall

The Blackwell Science logo is a
trade mark of Blackwell Science Ltd,
registered at the United Kingdom
Trade Marks Registry

DISTRIBUTORS

Marston Book Services Ltd
PO Box 269
Abingdon, Oxon OX14 4YN
(*Orders*: Tel: 01235 465500
            Fax: 01235 465555)

USA
Blackwell Science, Inc.
Commerce Place
350 Main Street
Malden, MA 02148 5018
(*Orders*: Tel: 800 759 6102
                  781 388 8250
           Fax: 781 388 8255)

Canada
Copp Clark Professional
200 Adelaide St West, 3rd Floor
Toronto, Ontario M5H 1W7
(*Orders*: Tel: 416 597-1616
                  800 815-9417
           Fax: 416 597-1617)

Australia
Blackwell Science Pty Ltd
54 University Street
Carlton, Victoria 3053
(*Orders*: Tel: 3 9347 0300
            Fax: 3 9347 5001)

A catalogue record for this title
is available from the British Library

ISBN 0-632-04953-7

Library of Congress
Cataloging-in-publication Data

Gaston, Kevin J.
    Biodiversity: an introduction/
  Kevin J. Gaston and John I. Spicer.
        p.   cm.
    Includes bibliographical references
  and index.
    ISBN 0–632–04953–7
    1. Biological diversity.
    I. Spicer, John I.   II. Title.
  QH541.15.B56G37   1998
  333.95'11—dc21

Contents

Preface, ix

1 What is biodiversity?
1.1 Introduction, 1
1.2 Biodiversity and the Rio conference, 1
1.3 Elements of biodiversity, 2
1.4 Measuring biodiversity, 5
    1.4.1 Number and difference, 5
    1.4.2 Value, 6
    1.4.3 Surrogacy, 6
    1.4.4 Species richness as a common currency, 10
1.5 Summary, 11
1.6 Making connections, 12
1.7 References, 12
1.8 Further reading, 12
    1.8.1 For this chapter, 12
    1.8.2 General texts on biodiversity, 13
    1.8.3 Surfing the World Wide Web (WWW), 13

2 Biodiversity through time
2.1 Introduction, 15
2.2 The fossil record, 15
2.3 A brief history of biodiversity, 16
    2.3.1 Principal features, from the beginning to the present day, 16
    2.3.2 Diversification, 23
    2.3.3 Extinction, 26
2.4 How many extant species are there? 28
2.5 Recent and future extinctions, 32
    2.5.1 Species losses, 32
    2.5.2 Population losses and declines, 35
    2.5.3 Biodiversity in crisis, 37
2.6 Ups as well as downs? 37
2.7 Summary, 38
2.8 Making connections, 38
2.9 References, 38
2.10 Further reading, 41

3 Mapping biodiversity
  3.1 Introduction, 43
  3.2 Issues of scale, 44
      3.2.1 Species–area relationships, 44
      3.2.2 Local–regional diversity relationships, 46
  3.3 Extremes of high and low diversity, 47
      3.3.1 Biological realms, 47
      3.3.2 Biogeographic regions, 48
      3.3.3 Provinces, 50
      3.3.4 Endemism, 51
  3.4 Gradients in biodiversity, 53
      3.4.1 Latitudinal gradients in diversity, 53
      3.4.2 Altitudinal and depth gradients in diversity, 58
      3.4.3 Peninsulas and bays, 62
  3.5 Diversity and environmental variables, 62
  3.6 Congruence, 65
  3.7 Conclusion, 67
  3.8 Summary, 69
  3.9 Making connections, 70
  3.10 References, 70
  3.11 Further reading, 75

4 Does biodiversity matter?
  4.1 Introduction, 76
  4.2 Use value, 76
      4.2.1 Direct use value, 76
      4.2.2 Indirect use value, 80
  4.3 Non-use value, 81
      4.3.1 Option and bequest values, 81
      4.3.2 Intrinsic value, 82
  4.4 How much biodiversity? 84
      4.4.1 Only one Earth, 84
      4.4.2 How much is enough? 85
  4.5 Conclusion, 87
  4.6 Summary, 87
  4.7 Making connections, 88
  4.8 References, 88
  4.9 Further reading, 89

5 Maintaining biodiversity
  5.1 Introduction, 91
  5.2 The scale of the human enterprise, 92
  5.3 The Convention on Biological Diversity, 94
      5.3.1 Article 1. Objectives of the Convention, 96
      5.3.2 Article 6. General Measures for Conservation and
          Sustainable Use, 96

5.3.3 Article 7. Identification and Monitoring, 97
5.3.4 Article 8. *In-situ* Conservation, 98
5.3.5 Article 9. *Ex-situ* Conservation, 101
5.3.6 Article 10. Sustainable Use of Components of
        Biological Diversity, 103
5.3.7 Article 11. Incentive Measures, 104
5.4 Responses to the Convention, 105
5.5 Conclusions, 105
5.6 Summary, 106
5.7 References, 107
5.8 Further reading, 108
5.8.1 The Convention on Biological Diversity, 108
5.8.2 Conserving biodiversity, 108
5.8.3 Conservation biology, 109

Index, 111

Preface

Like most prefaces, this was the last part of the book to be completed. As well as introducing the book to you, the reader, it also gives us, the authors, the chance to look back—to what motivated us to clear space in our calendars to write the pages that follow—and to reflect on the extent to which the emergent text has been modified by subsequent events.

Put simply, we felt that at the time there was no one simple, short introduction to the topic of biodiversity. So the question arose, if one was genuinely interested in getting to grips with the central tenets of this topic, without specialist help, where would one start and what would those most important 'central tenets' be? So the idea for *Biodiversity: An Introduction* came into being; a simple and short introductory text. In the course of writing the book we have repeatedly been challenged by arguments and expressed views (mostly good-humoured), from many different quarters, that 'biodiversity' is simply a buzz-word, a trendy topic with little real scientific credibility or value. Conversely, as we have become immersed (at times almost literally) in the vast scientific litera-ture on biodiversity we have found that the best of that literature conveys something of a rather different message—there are serious attempts to quantify and describe biodiversity in a scientific way, and we do not merely have to rely on 'just so' stories for our knowledge base. Consequently, our approach to introducing biodiversity—how it arose, where it occurs, why it is important, and what should be done to maintain it—has where ever possible been a quan-titative one.

Our approach has also been to cover as much ground as possible in as few pages as feasible. This reflects our belief that what is required at this time is a wide overview—putting the fine detail to one side and focusing on the 'big picture'. By definition this results in what may appear to be a superficial treat-ment of many important subjects. Consequently, each chapter also serves as an introduction to the primary literature, and concludes with an annotated 'Further reading' section. Similarly, the notion of biodiversity, and the science associated with it, lends itself to debate and discussion and so we have incorpo-rated such features not directly into the text, but into a number of learning exercises, scattered throughout. Here the reader is given the opportunity to explore, discuss, and debate key areas of controversy or interest, and come to their own conclusions (either entirely independently, or preferably as part of a

Convention on Biological Diversity. On one level the Convention document provides a helpful 'literary device', but its use also reflects a growing belief on our own part that, whatever its faults and short comings, this document provides one of the few opportunities realistically to address the current biodiversity crisis.

While the book as it is written takes a quantitative approach to biodiversity, this should not be taken to mean that we, as authors, only value aspects of biodiversity we can quantify. Even within the period of writing this book we have been struck by the sight of many of our students captivated by the sheer variety and wealth of life encapsulated within an intertidal pool and even (albeit as preserved specimens) within the departmental museum, and discussing with feeling how best to conserve biodiversity. What is more, one of the authors has stood amongst expectant crowds on a cold winter's day waiting for a glimpse of a rare bird, while the other, in the same month, has waded fully clothed into a chest-high pool to see some solitary corals.

The UK Biodiversity Action Plan captures this spirit neatly when it says that 'Biodiversity is life around us. It is a wonder and a delight. . . .' However, it goes on to say that 'it is also a concern and a responsibility'. With good quantitative understanding of biodiversity we will be better placed to meet this challenge.

Any measure of success we have had in achieving our goal for this volume owes much to the wisdom, insights, and editorial corrections of colleagues, friends and family. In particular, we acknowledge with gratitude the assistance of Tim Blackburn, Steven Chown, Sian Gaston, Ian Gauld, Len Hill, Bill Kunin, Phil Warren and Paul Williams. All kindly undertook to read and comment on part or all of the book manuscript. We also thank many of our students, particularly those taking module APS215 and those in our tutorial groups, who have (often unwittingly) test driven sections of the text and most of the exercises. Ian Sherman and Anna Woodford of Blackwell Science have, as ever, been a pleasure to work with. Sian and Fiona provided enthusiasm and encouragement in equal measure. Finally, we dedicate this book to Megan, Ben, Ethan and Ellie, who constantly remind us of the simple joys of observing biodiversity in the world about us.

K.J.G. & J.I.S.
Sheffield

1 What is biodiversity?

1.1 Introduction

The living world about us is infinitely, and gloriously, complex. The classic, and most familiar, examples of this observation are provided by a lowland tropical forest or perhaps by a coral reef. However, Charles Darwin also made the point with reference to a 'tangled bank', and with charity it might even be made of the denizens of a drop from a puddle of rainwater! Artists, composers, philosophers, poets and theologians, as well as biologists, have all endeavoured in their own way to encapsulate, to express, and frequently to celebrate, the complexity we encounter in the living world around us. As their efforts testify, exemplified by the countless books and articles written on this subject, such a task is far from simple.

The notion of biodiversity (a contraction of 'biological diversity') is one attempt to capture the complexity of life, and in so doing to further an understanding of it and hopefully of its maintenance. The term has become a familiar feature of our news programmes and papers. Importance is attached to it by environmental groups, political decision-makers, economists and ordinary citizens alike. But what does the term 'biodiversity' actually stand for, and what do we mean when we talk about describing or maintaining biodiversity? In this chapter we take arguably the most important definition of the term 'biodiversity' and ask how we can best assess biodiversity, given both very real practical constraints and the differences we, as a global society, place on various aspects of biodiversity.

1.2 Biodiversity and the Rio conference

There were, at the last count, more than 12 formal published definitions of the terms 'biological diversity' and 'biodiversity' (we use the two terms interchangeably). Of these, perhaps the most important and far-reaching is that contained within the Convention on Biological Diversity (the definition is provided in Article 2). This landmark treaty was signed by more than 150 nations on 5 June 1992 at the United Nations Conference on Environment and Development, held in Rio de Janeiro, and came into force approximately 18 months later (we subsequently refer to it simply as 'the Convention', although elsewhere you will commonly find it referred to by its acronym, CBD).

The Convention states that:

> *'Biological diversity' means the variability among living organisms from all sources including,* inter alia, *terrestrial, marine and other aquatic ecosystems and the ecological complexes of which they are part; this includes diversity within species, between species and of ecosystems.*

Put more colloquially, biodiversity is the variety of life, in *all* its many manifestations. It encompasses all forms, levels and combinations of natural variation and thus serves as a broad unifying concept.

For the purposes of our exploration of biodiversity we will amplify the full definition from the Convention in only one way. At present it does not obviously take into account the tremendous variety of biological life that occurred in the past, as preserved in the fossil record. However, we want to trace the origins of present-day biodiversity and this will necessitate delving into the past (see Chapter 2). To avoid any possible confusion therefore, we explicitly interpret the definition to embrace the variability of all organisms that have ever lived, and not simply those that are presently extant.

The actual definition of biodiversity, as given above, is neutral with regard to any importance it may be perceived to have. The Convention is, in contrast, far from a neutral document, as amply revealed by its objectives (Article 1):

> *. . . the conservation of biological diversity, the sustainable use of its components and the fair and equitable sharing of the benefits arising out of the utilization of genetic resources, including by appropriate access to genetic resources and by appropriate transfer of relevant technologies, taking into account all rights over those resources and to technologies, and by appropriate funding.*

Likewise, much of the usage of the term 'biodiversity', particularly in the media and by some politicians, is value laden. It carries with it connotations that biodiversity is a good thing *per se*, that its loss is bad, and that something should be done to maintain it. Consequently it is important to recognize that there is rather more to use of the term than a formal definition in the Convention (or for that matter elsewhere) and its application often reveals just as much about the values of the person using it (see Section 1.4.2 & Chapter 4). This should always be kept in mind when interpreting what is being said about biodiversity.

1.3 Elements of biodiversity

As stated in the definition of biodiversity, the variety of life is expressed in a multiplicity of ways. We can begin to make some sense of this variety by distinguishing between different key elements. These elements are the basic building blocks of biodiversity. They can be divided into three groups: genetic diversity, organismal diversity and ecological diversity (Table 1.1). Although presented separately the groups are intimately linked, and in some cases share elements in common (e.g. populations appear in all three).

Table 1.1 Elements of biodiversity. (After Heywood & Baste 1995.)

Ecological diversity	Genetic diversity	Organismal diversity
Biomes		Kingdoms
Bioregions		Phyla
Landscapes		Families
Ecosystems		Genera
Habitats		Species
Niches		Subspecies
Populations	Populations	Populations
	Individuals	Individuals
	Chromosomes	
	Genes	
	Nucleotides	

Some of these elements are more readily, and more consistently, defined than are others. When we consider genetic diversity, nucleotides, genes and chromosomes are discrete, readily recognizable and comparative units. Things are not quite so straightforward and neat when we move up to individuals and populations, with complications being introduced by, for example, the existence of clonal organisms and difficulties in identifying the spatial limits to populations. When we come to organismal diversity, most of the elements are perhaps best viewed foremost simply as convenient human constructs. For instance, debate persists over exactly how many kingdoms there should be, and recently a 'natural' three-domain classification has been proposed (see Box 2.1). When we refer to an order of clams and an order of mammals, for example, we are not necessarily comparing like with like, although within both clams and mammals orders may be broadly comparable. Even the reality and recognition of species, long considered one of the few biologically meaningful elements, has been a recurrent theme of debate for many decades, and a broad range of opinions and viewpoints have been voiced (Box 1.1). Finally, and perhaps most problematic, is exactly how we define the various elements of ecological diversity. In most cases these elements constitute useful ways of breaking up continua of phenomena. However, they are difficult to distinguish without recourse to what ultimately constitute some essentially arbitrary rules. For example, whilst it is helpful to be able to label different habitat types, it is not always obvious precisely where one ends and another begins.

Does the Convention help us to define the elements of biodiversity? Unfortunately the answer is no. The Convention refers to them, but explicitly defines very few. None the less, while it is true that many of the elements may be difficult to define rigorously, and in some cases may have no biological reality, they

> **Box 1.1** What constitutes a species?
>
> If we are to use species richness as a common currency for investigating biodiversity, it is prudent to ask what we mean when we refer to a species. Although this taxonomic unit is fundamental to all of biology, it is actually quite difficult to give a straightforward answer to this question. There are at least seven different concepts that address what a species is (based on Bisby 1995).
>
> 1 *Biological species*: group of interbreeding natural populations that do not successfully mate or reproduce with other such groups (and, some would add, which occupy a specific niche).
>
> 2 *Cohesion species*: smallest group of cohesive individuals that share intrinsic cohesive mechanisms (e.g. interbreeding ability, niche).
>
> 3 *Ecological species*: a lineage which occupies an adaptive zone different in some way from that of any other lineage in its range and which evolves separately from all lineages outside its range.
>
> 4 *Evolutionary species*: a single lineage of ancestor–descendant populations distinct from other such lineages and which has its own evolutionary tendencies and historical fate.
>
> 5 *Morphological species*: smallest natural populations permanently separated from each other by a distinct discontinuity in heritable characteristics (e.g. morphology, behaviour, biochemistry).
>
> 6 *Phylogenetic species*: smallest group of organisms that is diagnosably distinct from other such clusters and within which there is parental pattern of ancestry and descent.
>
> 7 *Recognition species*: a group of organisms that recognize each other for the purpose of mating and fertilization.
>
> As the vast majority of species have been, and are still being, described using differences in morphological characteristics, references to species richness more often than not concern 'morphological' species richness. Fortunately, this method of defining a species continues to be relatively effective for most needs.

remain useful and important tools for thinking about and studying biodiversity.

The elements of biodiversity, however defined, are not independent. Within each of the three groups of genetic, organismal and ecological diversity, the elements of biodiversity can be viewed as forming nested hierarchies (see Table 1.1, which serves also to render the complexity of biodiversity more tractable). Thus within genetic diversity, populations are constituted of individuals, each individual has a complement of chromosomes, these chromosomes comprise numbers of genes, and genes are constructed from nucleotides. Likewise, within organismal diversity, kingdoms, phyla, families, genera, species, subspecies, populations and individuals form a nested sequence, in which all elements at lower levels belong to one example of each of the elements at higher levels. This hierarchical organization of biodiversity reflects one of the central organizing principles of modern biology.

Whether any one element of biodiversity, from each or all of the three groups, can be regarded in some way as the most fundamental, essential or even natural is a contentious issue. For some, genes are the basic unit of life. However, in practice, it is often the species that is treated as the most fundamental element of biodiversity. Whether or not such an approach is useful, never mind correct, we will return to later (see Section 1.4.4).

1.4 Measuring biodiversity

1.4.1 Number and difference

Devising ways of estimating biodiversity quantitatively remains an unsolved problem. Maddox (1994)

For many purposes the concept of biodiversity is useful in its own right, as it can provide us with a valuable shorthand expression for what we have seen to be a very complex and important phenomenon. However, to be of more general applicability, we need to be able to measure biodiversity—to quantify it in some way. Only then can we address such fundamental questions as how biodiversity has changed through time, where it occurs and how we can maintain it.

Measures of diversity in general, and not solely of biodiversity, are commonly found in basic ecological texts. Essentially, many of these measures have two components: the number of entities and the degree of difference (dissimilarity) between those entities. For example, species richness (the number of species) places emphasis on the number of elements. However, weighting each of these species by, say, the numbers of individuals, would be one way of incorporating a metric of the differences between them into a measure. In the case of biodiversity the entities are one of its elements.

In measuring biodiversity, the breadth of ways in which differences can be expressed is potentially infinite. Think, for example, of the ways in which one could discriminate between two species. These might include facets of their biochemistry, biogeography, ecology, genetics, morphology, physiology or even the ecological role they play in a particular community (shredder, decomposer, predator, etc.) (cf. Box 1.1). As a result of the variety of elements of biodiversity, and of differences between them, there is no single all-embracing measure of biodiversity — nor will there ever be one! This means that it is impossible to state categorically what is *the* biodiversity of an area or of a group of organisms. Instead, only measures of certain components can be obtained, and even then such measures are only appropriate for restricted purposes.

Whilst one may feel uncomfortable with this notion, it is important to realize that it also applies, though perhaps not so obviously, in making many other concepts operational. For example, the topic of complex systems is attracting wide interest across a spectrum of fields of research (including physics), but there is no single measure of complexity (or simplicity for that matter). Instead there are many measures, none necessarily any more correct than the other and which quantify rather different things. To take an example closer to home, the concept of body size is widely utilized in biology. We can recognize that relationships exist between body size and latitude (the biggest butterflies are found in the tropics) or between body size and abundance (elephants are rarer than many mice). And yet there is no such thing as *the* body size of an organism. Rather, size can be (and is) expressed in a variety of ways, none of which has any obvious logical precedence. Consider, for example, two individuals similar in body mass but differing in linear dimensions. Which is the larger?

1.4.2 Value

We have mentioned before that measures of biodiversity are commonly used as bases for making decisions about conservation action or for planning more generally. It should now be clear that the choice of measure employed for evaluating biodiversity may not be neutral with regard to the outcome of such decisions. Different measures may suggest different answers. Moreover, it is important to remember that concentration on a particular element of biodiversity essentially places differential value on that facet of the variety of life (see Chapter 4). Both what you are measuring and how you are measuring it reveals something about what you most value. For example, if we use measures of ecological diversity as a basis for decision-making, this implies that this is the dimension of biodiversity that is of most importance to us. With this in mind we now consider the possibility of getting a handle on biodiversity in a quantitative way using surrogate measures.

1.4.3 Surrogacy

Not only is there no single measure of biodiversity, but even when you have chosen which facet to quantify it is rarely possible in practice to measure biodiversity in the way one might ideally like. One solution to this problem is to find surrogate measures, which are correlates of the actual measures desired but which can more readily be quantified. Perhaps one of the best examples of a surrogate measure of biodiversity derives originally from research in palaeontology. Here, the vagaries of the fossil record, and problems concerning what constitutes a 'fossil species', often make it very difficult to work at the species level (cf. Section 2.2); too few species are fossilized and subsequently recovered for patterns in species richness to be usefully explored. The use of higher taxa, typically families, as surrogates for species potentially overcomes this problem (cf. Section 2.3). This approach has also been found to work well more generally for present biota as well as fossils (Gaston & Williams 1993; Williams &

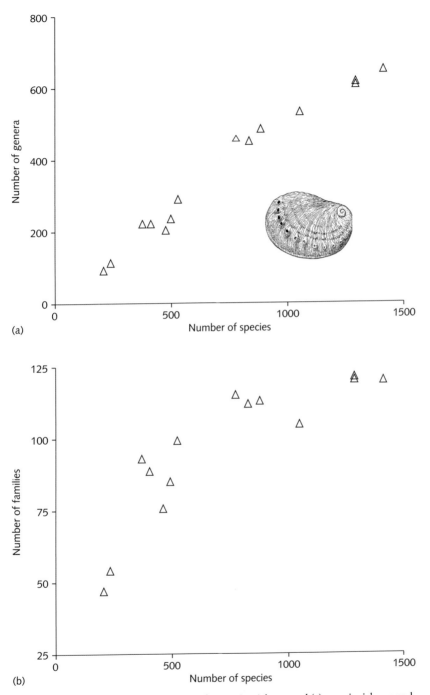

Figure 1.1 Relationships between present-day species richness and (a) generic richness and (b) family richness of eastern Pacific benthic molluscs in different latitudinal bands. (After Roy *et al.* 1996.)

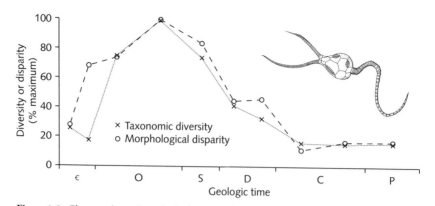

Figure 1.2 Changes through geological time in the numbers of higher taxa and in the morphological diversity (measured as average morphological dissimilarity among contemporaneous species) of a group of Palaeozoic marine invertebrates, the Blastozoa. (After Foote 1996.)

Gaston 1994). For example, the richness of both genera and families of eastern Pacific benthic molluscs has been found to increase with their species richness (Fig. 1.1). However, this approach should be applied with caution, because there are cases where numbers of higher taxa do not function effectively as surrogates for numbers of species, particularly when the ratio of numbers of higher taxa to numbers of species becomes extremely low.

Numbers of higher taxa may also act as surrogates for other facets of biodiversity. Thus, again using a palaeontological example, Foote (1996) has shown that taxonomic diversity is a good surrogate for morphological diversity in fossil blastozoan echinoderms (Fig. 1.2). The change in the number of orders through geological time is a reasonable predictor of morphological diversity at any given time. In general there is a trade-off between the precision of a surrogate and the ease with which it can be measured. Surrogates that reflect the desired quantity particularly well tend, unsurprisingly, to be difficult to measure.

Some surrogates inevitably capture more of the patterns of number and difference than do others. One example is phylogenetic diversity, a measure of how closely related, in evolutionary terms, are the species in an assemblage (Vane-Wright *et al.* 1991; Nixon & Wheeler 1992; Williams & Humphries 1996). In general, patterns of difference among species are most likely to be congruent with the pattern of their genealogical relationships through genetic inheritance. Close relatives, say two species of mouse, are, on average, more likely to share similar biologies than are distant relatives, say a species of mouse and a species of shrimp. The level of phylogenetic diversity thus tends to capture not only the degree of relationship but also the degree of difference in many other characteristics. There are, of course, some significant exceptions (typified by strongly divergent evolution of close relatives and strongly convergent evolution of distant relatives). For example, some closely related organisms are very different in their morphology: the crustacean family Cirripedia encompasses not only the barnacles proper, but also a strange-looking parasitic group, the rhizocephalans. In these parasitic barnacles the 'body' is ramified through the whole crab host with the only external (to the host) feature being a very large reproductive sac. None the less, on average, the principle holds.

Perhaps chief amongst surrogates is species richness (number of species). In general, as long as the numbers involved are at least moderate, it could be argued that greater numbers of species tend to embody more genetic diversity (in the form of a greater diversity of genes through to populations), more organismal diversity (in the form of greater numbers of individuals through to higher taxa) and greater ecological diversity (from representatives of more niches and habitats through to more biomes). For example, species richness is

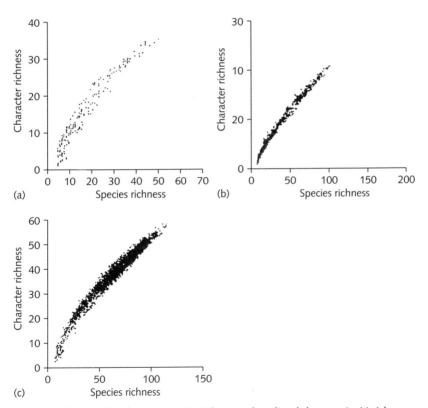

Figure 1.3 Relationships between species richness and predicted character (trait) richness: (a) at the global scale for species of bumble-bees among 611 000 km² grid cells; (b) at the regional scale for species of selected plant taxa among 1° × 1° grid cells in the Neotropics; and (c) at the national scale for species of breeding birds among 10 × 10 km grid cells in Britain. (After Williams & Humphries 1996.)

positively correlated with numbers of higher taxa (see Fig. 1.1) and character richness (Fig. 1.3).

1.4.4 Species richness as a common currency

As we have seen, whilst biodiversity can be measured in a whole host of ways, in practice it tends to be measured in terms of species richness (see Box 1.1). There are several reasons why this is so.

1 It is a good surrogate as it acts as an 'integrator' of many facets of differences in biodiversity.

2 It is frequently measurable in practice (though there are some significant complications, particularly associated with dependence on sample size).

3 A substantial amount of information already exists on patterns in species richness (and more can readily be extracted from existing museum collections and associated literature, etc.).

The limitation of species richness as a measure of biodiversity has frequently been illustrated with reference to the issue of whether an assemblage of a small number of closely related species, say our earlier two species of mouse, is more or less biodiverse than an equivalent-sized assemblage of more distantly related species, say our earlier species of mouse and species of shrimp (see Section 1.4.3). As we have noted, while the latter assemblage would intuitively seem to be the more diverse, in terms of species richness the assemblages are equally diverse. The extent to which this is a weakness of using species richness as a measure of biodiversity depends perhaps less on the outcomes of such simple scenarios than on scenarios more typical of studies of biodiversity, which commonly involve assemblages numbering at least tens if not hundreds or thousands of species. Here, it seems that species richness is often strongly positively correlated with many other measures of biodiversity (Gaston 1996).

Species richness has, in some sense, become the common currency of the study of biodiversity. If we wish to explore and discuss the origin, patterns and maintenance of biodiversity we do require a common currency to make the task manageable. This is particularly important in such a small book as this. So throughout the rest of this book we will essentially treat species richness as equivalent to biodiversity, notwithstanding the facts that it remains only one among many measures and retains some significant and important limitations. In so doing, we do not wish to imply that the problems associated with using this one measure are either trivial or unimportant. However, progress can be made using it, provided we remain alert to its limitations. Moreover, we will also have recourse to use some other measures, and the broader dimensions of biodiversity must always be borne in mind.

Going Further

As part of a group
• Debate the statement: 'When measuring biodiversity there is no realistic alternative to species richness at present'.
• Split into two groups each comprising three to five members. Appoint a chairperson (he or she can be a member of one of the groups) who is responsible for making sure that what is supposed to happen does happen.
• Members of each group should then read Sections 1.3 and 1.4 (again), together with the references given below. Relevant information should be extracted in order that one group can prepare arguments for, and the other group against, the motion. The amount of time to be spent reading and preparing a presentation is negotiable but should be agreed at the outset.
• Each group will have 10 minutes to present their argument to the opposing group and must not be interrupted. The method of presentation is up to the group members but must be appropriate and engaging. The chairperson will be responsible for both groups keeping to time.
• After the formal presentations the chairperson should give both groups a chance to cross-examine and challenge each other, but one at a time. This should take, in total, no more than 20 minutes. At the end of this time, the chairperson should ask all those involved to see if a consensus view has emerged/is emerging (15 minutes).
• The session should end with the chairperson attempting to summarize points of agreement as well as unresolved difficulties (5 minutes).

References
Gaston (1996), Mallet (1996), Martinez (1996), Williams and Humphries (1996)

1.5 Summary

1 Biodiversity is the variety of life, in all its manifestations.
2 Key elements of this variety can be recognized, comprising three nested hierarchies of genetic, organismal and ecological diversity.
3 There is no single overall measure of biodiversity; rather there are multiple measures of different facets.
4 The measure of biodiversity chosen may reveal something about the values of the investigator.
5 As many desired measures of biodiversity are difficult or impractical to obtain, surrogate measures (correlates of the actual measures desired) may have to be used instead. Some surrogates take into account more of the patterns of numbers and difference in biodiversity in which we are interested than do others.
6 Despite its significant limitations species richness is one of the most commonly used surrogates as it 'integrates' many different facets (genetic, organismal, ecological) of biodiversity. It is relatively easy to measure and, despite significant limitations, has to some extent become the common currency of the study of biodiversity.

1.6 Making connections

Having agreed on what we mean by 'biodiversity' and decided on the most convenient way of measuring it for present purposes, in the next chapter we will look at how changes in biodiversity through geological time, and also in recent times, have resulted in the present-day diversity of life on Earth.

1.7 References

Bisby, F.A. (1995) Characterization of biodiversity. In: *Global Biodiversity Assessment* (ed. V.H. Heywood), pp. 21–106. Cambridge University Press, Cambridge.

Foote, M. (1996) Perspective: evolutionary patterns in the fossil record. *Evolution* 50, 1–11.

Gaston, K.J. (1996) Species richness: measure and measurement. In: *Biodiversity: a Biology of Numbers and Difference* (ed. K.J. Gaston), pp. 77–113. Blackwell Science, Oxford.

Gaston, K.J. & Williams, P.H. (1993) Mapping the world's species – the higher taxon approach. *Biodiversity Letters* 1, 2–8.

Heywood, V.H. & Baste, I. (1995) Introduction. In: *Global Biodiversity Assessment* (ed. V.H. Heywood), pp. 1–19. Cambridge University Press, Cambridge.

Maddox, J. (1994) Frontiers of ignorance. *Nature* 372, 11–36.

Mallet, J. (1996) The genetics of biological diversity: from varieties to species. In: *Biodiversity: a Biology of Numbers and Difference* (ed. K.J. Gaston), pp. 13–53. Blackwell Science, Oxford.

Martinez, N.D. (1996) Defining and measuring functional aspects of biodiversity. In: *Biodiversity: a Biology of Numbers and Difference* (ed. K.J. Gaston), pp. 114–148. Blackwell Science, Oxford.

Nixon, K.C. & Wheeler, Q.D. (1992) Measures of phylogenetic diversity. In: *Extinction and Phylogeny* (eds M.J. Novacek & Q.D. Wheeler), pp. 216–234. Columbia University Press, New York.

Roy, K., Jablonski, D. & Valentine, J.W. (1996) Higher taxa in biodiversity studies: patterns from eastern Pacific marine molluscs. *Philosophical Transactions of the Royal Society of London Series B* 351, 1605–1613.

Vane-Wright, R.I., Humphries, C.J. & Williams, P.H. (1991) What to protect? Systematics and the agony of choice. *Biological Conservation* 55, 235–254.

Williams, P.H. & Gaston, K.J. (1994) Measuring more of biodiversity: can higher-taxon richness predict wholesale species richness? *Biological Conservation* 67, 211–217.

Williams, P.H. & Humphries, C.J. (1996) Comparing character diversity among biotas. In: *Biodiversity: a Biology of Numbers and Difference* (ed. K.J. Gaston), pp. 54–76. Blackwell Science, Oxford.

1.8 Further reading

1.8.1 For this chapter

Claridge, M.F., Dawah, H.A. & Wilson, M.R. (eds) (1997) *Species: the Units of Biodiversity.* Chapman & Hall, London.

Gaston, K.J. (1996) What is biodiversity? In: *Biodiversity: a Biology of Numbers and Difference* (ed. K.J. Gaston), pp. 1–9. Blackwell Science, Oxford. (*Takes a different view from the one proffered here, distinguishing between biodiversity as a concept, a measurable entity and a social/political construct.*)

Gaston, K.J. (1998) Biodiversity. In: *Conservation Science and Action* (ed. W.J. Sutherland). Blackwell Science, Oxford. (*An attempt at a one chapter summary of some of the primary issues in the study of biodiversity.*)

Hawksworth, D.L. (ed.) (1995) *Biodiversity: Measurement and Estimation.* Chapman & Hall, London. (*An important, if somewhat eclectic, set of papers.*)

Magurran, A.E. (1988) *Ecological Diversity and its Measurement.* Croom Helm, London. (*Lucid review and a good point of entry into this field.*)

Noss, R.F. (1990) Indicators for monitoring biodiversity: a hierarchical approach. *Conservation Biology* 4, 355–364. (*Distinguishes an alternative hierarchical organization to biodiversity, based on composition, structure and function.*)

1.8.2 General texts on biodiversity

Dobson, A.P. (1996) *Conservation and Biodiversity.* Scientific American, New York. (*Beautifully produced and reasonably comprehensive, with a good bibliography; very accessible.*)

Gaston, K.J. (ed.) (1996) *Biodiversity: a Biology of Numbers and Difference.* Blackwell Science, Oxford. (*A wide-ranging, but not comprehensive, examination of the measurement of, temporal and spatial patterns in, and the conservation and management of biodiversity.*)

Heywood, V.H. (ed.) (1995) *Global Biodiversity Assessment.* Cambridge University Press, Cambridge. (*Undoubtedly the most comprehensive review of the different facets of biodiversity, from characterization to economic importance. A formidable tome!*)

Huston, M.A. (1994) *Biological Diversity: the Coexistence of Species on Changing Landscapes.* Cambridge University Press, Cambridge. (*A very ecological perspective on biodiversity.*)

Reaka-Kudla, M.L., Wilson, D.E. & Wilson, E.O. (eds) (1997) *Biodiversity II: Understanding and Protecting our Biological Resources.* Joseph Henry Press, Washington, DC. (*A long-awaited sequel to Wilson & Peters (1988).*)

Solbrig, O.T. (ed.) (1991) *From Genes to Ecosystems: a Research Agenda for Biodiversity.* International Union of Biological Sciences (IUBS), Paris. (*Identifies some of the major issues to be addressed in the study of biodiversity.*)

Wilson, E.O. (1992) *The Diversity of Life.* Penguin Books, London. (*A popular, wide-ranging and very readable account by perhaps the most influential proponent of biodiversity.*)

Wilson, E.O. & Peter, F.M. (eds) (1988) *BioDiversity.* National Academy Press, Washington, DC. (*Where it all began? The 'milestone' volume that drew attention to the importance of biodiversity.*)

World Conservation Monitoring Centre (1992) *Global Biodiversity: Status of the Earth's Living Resources.* Chapman & Hall, London. (*A useful collation of essays and data.*)

World Conservation Monitoring Centre (Comp.) (1994) *Biodiversity Data Sourcebook.* World Conservation Press, Cambridge. (*An update and expansion of some of the information in the 1992 volume.*)

1.8.3 Surfing the World Wide Web (WWW)

'Biodiversity' on a search engine throws up a whole load of material, some useful and much not. To save you time there is a list of biodiversity WWW sites (http://www.biologie.uni-freiburg.de/data/zoology/riede/taxalinks.html), through which you can actually access the sites listed. However, there are three web sites that call for special mention.

The Convention on Biological Diversity and all of the material associated with it is accessible at http://www.unep.ch/biodiv.html.

The World Resources Institute (WRI) web site (http://www.wri.org/wri/biodiv/) is a valuable source of biodiversity facts and figures.

The World Conservation Monitoring Centre (WCMC) is an internationally recognized body for collation of information on conservation and sustainable use of biodiversity. Visitors to their web site (http://www.wcmc.org.uk/) will find good general information but also fairly detailed information in the form of statistics and maps, generated from their databases. These include details of protected areas, national biodiversity strategies and data on threatened species.

2 Biodiversity through time

2.1 Introduction

It is not unreasonable to suppose that an understanding of how biodiversity has arisen, and how it has changed in the past, may be important in interpreting its present and future structure. In this chapter, we consider the temporal dynamics of biodiversity, i.e. how biodiversity changes with time. We begin with a brief overview of the history of life and the principal historical patterns in the magnitude of biodiversity. We move then to consider how many extant species there are at present and the levels of recent and future extinctions. Finally we discuss the consequences for levels of biodiversity of the movements of species around the planet that result (intentionally or accidentally) from human activities.

2.2 The fossil record

> *If you wished to maintain that the sea shells in these mountains were engendered by nature with the aid of stars how would you explain that these stars succeeded in creating such shells of differing sizes, differing ages, and different species at the same spot?* Leonardo da Vinci (1452–1519)(from Wendt 1970, p. 22)

Our knowledge of the history of biodiversity derives principally from analyses of the fossil record. This is reasonable as 'speculations about the past, if they are not to be idle, must be based on what the past has left behind' (Ayer 1973). Much of our modern-day geological landscape owes its origins to past biodiversity. However, working with the fossil record is an important constraint for three reasons.

First, as recognized by Darwin when marshalling evidence for his theory of evolution, this record is far from perfect or even. Current estimates of the percentage of species leaving a fossil record range from less than one to, at most, a few per cent of those that have ever lived (e.g. Sepkoski 1992). Second, of this fossil record, only a fraction has actually been recovered. Third, the record, and that portion of it that has been recovered, is biased towards the more abundant and the more widespread species, and more towards some groups of organisms than others. For instance, soft-bodied organisms, such as some cnidarians (jellyfish, sea anemones) are rarely fossilized and are exceptional in the fossil record, whereas the number of individual brachiopod fossils has been estimated to be in the billions. Some of the major soft-bodied animal groupings have left no fossil remains: animals like the Platyhelminthes (flatworms, flukes and tapeworms). The fossil record for animals with hard body parts, such as the brachiopods, molluscs, echinoderms and vertebrates, while often much better, is still far from complete and not always representative: 95% of all fossil species are marine animals, while 85% of today's recorded plants and animals are terrestrial.

The paucity of the fossil record, even with regard to individual taxa, is well illustrated by a group that possesses hard body parts and is relatively well researched, having caught the attention and imagination of people of all ages and from all walks of life—the dinosaurs. Although something of the history of this group is known even to many primary school children, it remains based on a remarkably small window on the past. As of 1990, 900–1200 genera of dinosaurs were estimated to ever have lived (Dodson 1990). Of these, only 285 (336 species) were known from fossils and nearly half of these were known from only a single specimen; complete skulls and skeletons were known from only 20% of known genera.

While it is clear that the fossil record (and our 'picture' of it) is far from complete, in many different ways, it is still an invaluable 'pictorial' history of life on Earth, where many of the major events in that history have left their mark in, or on, the rocks. Notwithstanding its limitations, it is still possible to construct an understanding of changes in biodiversity through geological time using the fossil record. However, because of the constraints referred to above, throughout this chapter we will often have to make recourse to the temporal dynamics of numbers of higher taxa rather than species. This should not pose too much of a problem, for not only do numbers of higher taxa act as a surrogate for numbers of species (cf. Section 1.4.3) but it is also true that they act as a measure of biodiversity in their own right (see Table 1.1).

2.3 A brief history of biodiversity

2.3.1 Principal features, from the beginning to the present day

> *There is grandeur in this view of life, with its several powers, having been originally breathed by the creator into a few forms or into one; and that, whilst this planet has gone cycling on according to the fixed law of gravity, from so simple a beginning endless forms most beautiful and most wonderful have been, and are being evolved.* Darwin (1859)

Some of the 'major events' of life on Earth, together with their chronology, are presented in Table 2.1. Self-evidently, biodiversity has increased between its inception, estimated to be about 3.5–4.0 billion years ago (the Earth itself is thought to be more than 4.5 billion years old), and the present time, otherwise

Table 2.1 Geological eras and periods, and the major events associated with them. (After Schopf 1992.)

Era	Period	Mya.	Major events
Precambrian	P€	4500	Origin of life, first multicellular organisms
Palaeozoic	Cambrian (€)	550	All of the major phyla present in fossil record
	Ordovician (O)	500	First vertebrates (jawless fish)
	Silurian (S)	440	Colonization of land by plants and arthropods
	Devonian (D)	410	Diversification of teleosts (bony fish)
			First amphibians and insects
	Carboniferous (C)	360	Extensive forests of vascular plants, origin of reptiles, amphibians dominant
	Permian (P)	290	Mass extinction of marine invertebrates, origins of mammal-like reptiles and 'modern insects'
Mesozoic	Triassic (₮)	250	Origin and diversification of ruling reptiles, origin of mammals, gymnosperms dominant
	Jurassic (J)	210	Dominance of ruling reptiles and gymnosperms, origin of birds
	Cretaceous (K)	140	Origin of angiosperms (flowering plants), ruling reptiles and many invertebrate groups go extinct towards end of period
Cenozoic	Tertiary (T)	65	Diversification of mammals, birds, pollinating insects and angiosperms
			Late T/early Q: the zenith of biodiversity
	Quaternary (Q)	1.8	Origin of humankind

we would not see the wealth of organisms that we do today. At first its increase appears to have been very slow. One of the key innovations, which opened the door to a major increase in biodiversity, was the advent of multicellularity (i.e. the appearance of individual organisms being composed of numerous cells, differentiated for the performance of different functions). Multicellular organisms did not begin to diversify until perhaps 1.4 billion years ago, when nearly 60% of the history of life had already passed. Multicellular animals (metazoans) specifically did not begin to markedly diversify until approximately 600 million years (Ma) ago, by which time about 80% of the history of life had passed. None of these first fossil metazoans possessed any hard parts and most were no more than a few millimetres long. There are a few tantalizing glimpses of relatively large soft-bodied metazoans in late Precambrian (also known as Vendian) rocks, e.g. in the Ediacaran fauna in Australia, which has been referred to as comprising either ancestral metazoans or a parallel unsuccessful metazoan experiment.

It is only with the beginning of the Palaeozoic Era ('early life'), and in rocks of the Cambrian Period (550 Ma ago), that we see the 'sudden' appearance of the first sizeable metazoans with hard parts (as exemplified by the 'wonderful life' (Gould 1989) of the Burgess Shale fauna from Canada). Not only are the fossils plentiful but there are a bewildering array of different body plans present, some 'experimental' and relatively short-lived, but others surviving and remaining with us to the present day. In fact, by the end of the Cambrian we find that nearly all of today's major animal groupings (or phyla) are present in the fossil record. Gould (1989) suggests that anatomical diversity reached a maximum around this time. The colonization of land by animals and plants (440 Ma ago), and their subsequent diversification, lagged far behind the emergence of multicellular organisms in the oceans. So animal life has gone from a position of relatively few species encompassing many different body plans in the Cambrian ('early experimentation . . .'), to the present day where we see considerably more species but noticeably fewer body plans ('. . . and later standardization') (Gould 1989).

A list of all of the present-day phyla is presented in Box 2.1. While the list is impressive the distribution of members between these groupings is heavily skewed, with most being encompassed by relatively few phyla. To take one

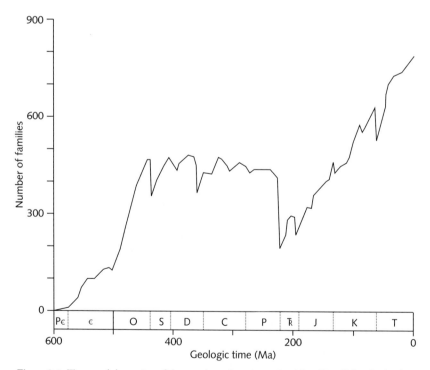

Figure 2.1 Temporal dynamics of the number of marine animal families. (After Sepkoski 1992.)

Box 2.1 A survey of present-day phyla

The currently accepted classification of life recognizes five kingdoms (Whittaker & Margulis 1978; Margulis *et al.* 1994), based on cell, organizational (unicellular/multicellular, solitary/colonial) and nutritional type, although a more 'natural' three-domain system has recently been proposed (Woese *et al.* 1990). Woese and his colleagues, comparing ribosomal RNA sequences, found that microbial life (i.e. prokaryotes: cell lacking nucleus) was composed of two separate groups, which they termed Eubacteria and Archaebacteria (see Williams & Embley 1996 for a recent review of microbial diversity in the context of the domains or kingdoms debate). All other life (i.e. eukaryotes: cell with nucleus) was placed into the third domain, the Eukaryota. Our list of present-day phyla has been constructed using various sources (Laverack & Dando 1987; Margulis *et al.* 1994; Funch & Kristensen 1995) and recognizing that it is only one possible list among many, as debate continues over exactly how many phyla there are. The main purpose of the list is that it allows the reader to survey at a glance the numbers, names and relative proportions of the different 'body plans' that constitute life on Earth; it also serves as an antidote to the notion that biodiversity is mainly concerned with plants and animals. It should be noted that this listing does not include the viruses, which are minute and mostly parasitic 'suborganisms' derived, it has been suggested, from the nuclear material of organisms.

Animalia
Acanthocephala
Annelida (true worms)
Arthropoda (insects, spiders, lobsters)
Brachiopoda (lampshells)
Chaetognatha (arrow worms)
Chordata (sea squirts, vertebrates)
Cnidaria (jellyfish, corals, sea anemones)
Ctenophora (sea gooseberries)
Cycliophora
Echinodermata (sea urchins, starfish)
Echiura
Ectoprocta
Entoprocta
Gastrotricha
Gnathostomulida
Hemichordata (acorn worms)
Kinorhyncha
Loricifera
Mesozoa
Mollusca (snails, slugs, octopus)
Nematoda (round worms)
Nematomorpha

Protoctista (Greek: *proteros*, very first; *ctistes*, to build or establish)
Acrasea (slime moulds)
Actinopoda
Apicomplexa
Bacillariophyta (diatoms)
Chlorarachnida
Chlorophyta
Chrysophyta
Chytridiomycota
Ciliophora (ciliates)
Conjugaphyta
Cryptophyta
Dictostelida
Dinomastigota (dinoflagellates)
Ebridians
Ellobiopsida
Euglenida (some flagellates)
Eustigmatophyta
Glaurocystophyta
Granuloreticulosa (foraminiferans)

(Box 2.1 continued on p. 20.)

Box 2.1 *Continued.*

Animalia
Nemertina (bootlace worms)
Onychophora (velvet worms)
Pentastoma
Phoronida
Placozoa
Platyhelminthes (flatworms, tapeworms)
Pogonophora (beard worms)
Porifera (sponges)
Priapulida
Rotifera (wheel animals)
Siphuncula (peanut worms)
Tardigrada (water bears)
Vestimentifera (tubeworms)

Plantae (**Latin:** *planta*, **plant**)
Bryophyta (liverworts, mosses)
Coniferophyta (conifers)
Cycadophyta (cycads)
Filicinophyta (ferns)
Ginkgophyta (ginkgo)
Gnetophyta
Lycopodophyta (club mosses)
Magnoliophyta (flowering plants)
Psilophyta
Sphenophyta (horsetails)

Fungi (**Latin:** *fungus*, **fungi**)
Ascomycota
Basidiomycota
Deuteromycota
Mycophycophyta (lichens)
Zygomycota

Protoctista
Haplosporidia
Hyphochytriomycota
Karyoblastea
Labyrinthulomycota (slime nets)
Microspora
Myxozoa
Oomycota
Paramyxea
Phaeophyta (brown algae)
Myxomycota (slime moulds)
Plasmodiophoromycota
Prymnesiophyta
Raphidophyta
Rhizopoda (amoebae)
Rhodophyta (red algae)
Xanthophyta
Xenophyophora
Zoomastigina

Monera (Prokaryotae or bacteria)
  (Greek *pro*, before; *karyon*,
  nut or nucleus)
Actinobacteria
Aeroendospora
Anaerobic phototrophic bacteria
Aphragmabacteria
Chemoautotrophic bacteria
Cyanobacteria (blue-green algae)
Fermenting bacteria
Halophilic/thermoacidophilic
  bacteria
Methanocreatrices
Micrococci
Myxobacteria
N-fixing aerobic bacteria
Omnibacteria
Pseudomonas
Spirochaetae
Thiopneutes

striking example in the kingdom Animalia, the phylum Placozoa possesses but one species, compared with the Arthropoda where species number millions (cf. Section 2.4). Furthermore, although the oldest fossils are microbes, most of the

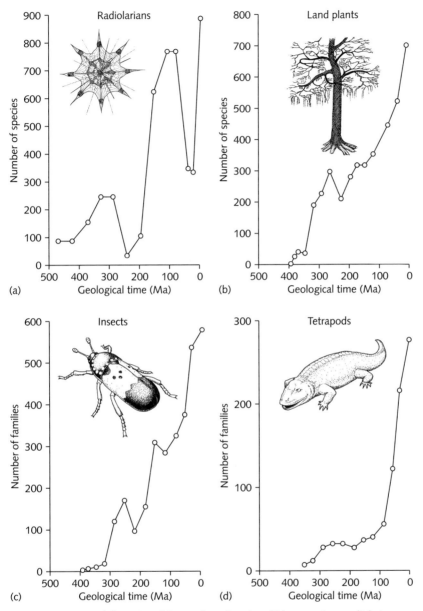

Figure 2.2 Temporal dynamics of the number of species of (a) protoctistan radiolarians and (b) land plants, and of the number of families of (c) insects and (d) tetrapod vertebrates. Data derived from Tappan and Loeblich (1973), Benson (1985), Niklas (1986) and Labandeira and Sepkoski (1993).

palaeontological focus, and arguably much of the modern-day biodiversity focus, has been on only two of the five kingdoms, the Animalia and the Plantae. Therefore, in what has gone before, as in so many other places in this book, most of our examples have had to be based solely on either plants or animals. This does not necessarily imply that these two kingdoms are any more important in terms of biodiversity than any of the others that exist.

Broadly speaking there were relatively few species during the Palaozoic and early Mesozoic eras (although this has been a matter of some controversy in the past; Signor 1990). However, starting just over 100 Ma ago there was a progressive and substantial increase in biodiversity that culminated at the end of the Tertiary and beginning of the Quaternary (Pleistocene) in there being more extant species and higher taxa of animals and plants (both marine and terrestrial) than at any time before or, indeed, since (Signor 1990). We are living in the Quaternary (Holocene) in a time of decreasing diversity, which is correlated with change in climate and the advent of organized and large-scale human activity.

Once diversification began to occur on a major scale (Cambrian to the present day) it was not, as we have seen above, continuous. Rather, there were periods of dramatic increase, interspersed by sometimes major setbacks or periods of relative stasis (or at least no marked directional trend in diversity). Consequently, the history of biodiversity is often presented as one of radiations and stabilizations, punctuated by mass extinctions (Signor 1990; Sepkoski 1992). This basic historical pattern is well illustrated by the pattern of temporal change in numbers of families of marine animals (Fig. 2.1). This exhibits three main phases of diversification (in the early Cambrian, in the Ordovician, and through the Mesozoic and Cenozoic), two main phases of approximate stabilization of diversity (in the mid to late Cambrian and through most of the Palaeozoic) and five major mass extinctions (end-Ordovician, Late Devonian, end-Permian, end-Triassic, end-Cretaceous). The same general pattern is also clearly seen in terrestrial systems, as illustrated by the history of diversity of plants, insects and tetrapods (Fig. 2.2).

For marine and terrestrial groups not only are the patterns of radiation, stabilization and extinction detectable, but superimposed upon them is a clear overall trend of increasing biodiversity (Fig. 2.2). The broad patterns of tempo-

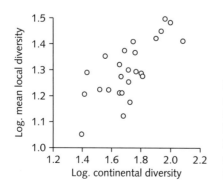

Figure 2.3 Relationship between local and continental species diversity through time for large mammalian carnivore and herbivore species in North America over the last 44 Ma (divided into 25 time intervals; each data point is for one time interval). (After Van Valkenburgh & Janis 1993.)

ral change are, to a first approximation, reflected both in global and regional biodiversity and in local biodiversity (Fig. 2.3). This is both interesting and informative, as it tells us that as biodiversity has increased on a global scale it has tended also to do the same locally (the alternative scenario would have been that biodiversity remained approximately constant locally, with the global increase having resulted solely from a growing differentiation between the occupants of different localities).

Given that there is a pattern of overall increase in biodiversity through time, the obvious question is why? The answer, quite simply, is that we do not know. A number of different factors have been suggested as effecting this increase: external factors, such as the break-up of the continents and their subsequent drift (increasing the differentiation between assemblages on different continents and in different ocean basins) and changing climatic conditions; and intrinsic factors, such as the occupation, through evolutionary time, of more and more of the potential niche space open to organisms (associated with evolutionary 'breakthroughs'), and perhaps finer subdivision of this space. The move on to land, for example, opened up many more opportunities for speciation than previously had existed.

2.3.2 Diversification

The overall pattern of diversification is not a product of synchronous changes in the biodiversity of all the component groups (Fig. 2.4). Rather, some groups

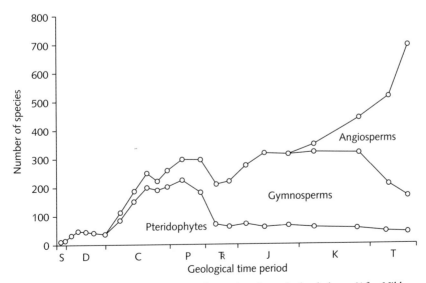

Figure 2.4 Changes in diversity over time for species of vascular land plants. (After Niklas 1986.)

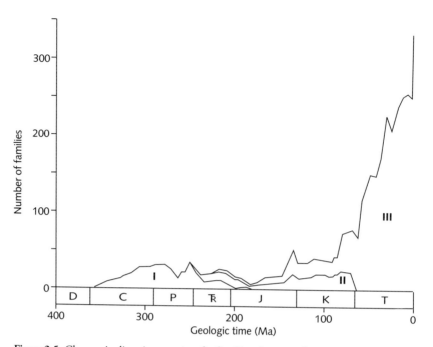

Figure 2.5 Changes in diversity over time for families of terrestrial tetrapods. I, early amphibians, early reptiles (anapsids), mammal-like reptiles. II, early diapsids (reptiles), dinosaurs, flying lizards. III, 'modern groups': amphibians (salamanders and frogs), reptiles (turtles, lizards, snakes and crocodiles), birds and mammals. (After Benson 1985.)

underwent differential diversification in particular time periods. This can clearly be seen with reference to the land plants and to the vertebrate tetrapods. Amongst the land plants, primitive vascular plants gave way to pteridophytes (ferns) and lycopsids (club mosses), which in turn gave way to a predominance of gymnosperms (spore-bearers), which finally were overtaken by the angiosperms (flower-bearers); there is some evidence that the angiosperms continue their diversification through the present. Amongst vertebrate tetrapods, the early amphibians and reptiles gave way to the ruling reptiles, which in turn gave way to the modern amphibians and reptiles, the birds and the mammals (Fig. 2.5). It is tempting to interpret these successions as cases of competitive replacement or improvement (with one group being driven out by the growing numbers of species of the next group). However, there is no reason that this interpretation need be correct, and the reasons for these patterns are almost always considerably more complex.

Notwithstanding the relatively large number of major body plans, or phyla (see Box 2.1), much of biodiversity is contributed by just a few groups of organisms, whilst most groups are simply not very diverse. This pattern is repeated at all taxonomic levels. Take for example the numbers of species in

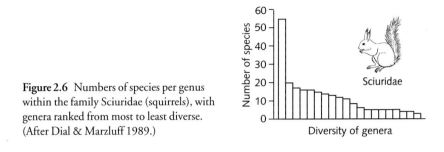

Figure 2.6 Numbers of species per genus within the family Sciuridae (squirrels), with genera ranked from most to least diverse. (After Dial & Marzluff 1989.)

each of the genera that comprise the family of mammals that contains the squirrels, the Sciuridae (Fig. 2.6). A large proportion of the total number of species in the family belongs to just one genus, *Sciurus*. The remaining species are distributed more or less equally between the remaining genera. We can see the same pattern on a much larger scale when we examine the extant flora and fauna. Most species of plants are angiosperms (>75%), most animals are insects (>80%) and more species of mammal belong to the rodents than to any other group (>40%). We even see this pattern in the Burgess Shale fauna, from Cambrian seas, where with 140 species being distributed between 124 genera many of the animals present are arthropods (Fig. 2.7). But what determines such differential patterns of diversification?

First, it is possible that this could merely be an artefact of the process of classifying organisms into groups, and may have no biological basis. There is little evidence that this is actually true, because the differences between many groups of organisms are clearly real and reflective of their evolutionary relationships; although curiously, humans do tend to organize sets of inanimate objects into a few large groups and many small ones.

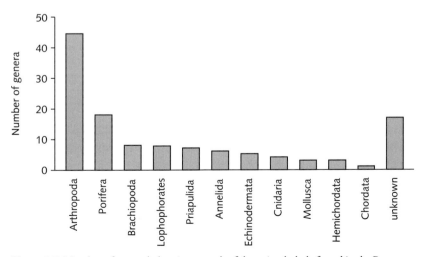

Figure 2.7 Number of genera belonging to each of the animal phyla found in the Burgess Shale (middle Cambrian) fauna. Data from Conway-Morris (1979).

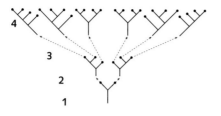

Figure 2.8 A diagrammatic representation of the possible routes by which lineage splitting leads from one ancestral species to four. (After Slowinski & Guyer 1989.)

Second, the patterns could simply be a matter of chance. Indeed, a pattern in which many groups have a few species and one or a few groups contain a high proportion of species is a likely product of a model of random speciation and extinction. Let us work through an example to illustrate this. Consider the circumstances in which lineage splitting leads from one ancestral species to four descendant species, and in which all branching points are dichotomous (Fig. 2.8). Initially an ancestral species splits to give two distinct species. Depending on which of these splits, two possible three-species outcomes exist; depending on which of these three species splits, six possible four-species outcomes may result. Of these patterns of phylogeny of four species, only 2/6 are symmetrical; an uneven distribution of species is the more likely outcome. Indeed, models of random speciation and extinction take us a long way towards understanding patterns of diversification, but are not always sufficient. Some groups still have disproportionately more species than would be expected by chance.

This leads us to the third possible reason for the observed pattern, which is that some groups have features that predispose them to diversify disproportionately. Thus it has been proposed that animal dispersal has promoted the diversification of some vascular plant groups, the ability to fly has promoted the diversification of some insect groups, and small body size has promoted the diversification of some bird groups. Such suggestions have proved much more difficult to test than was long supposed and there are many 'just-so' stories for why one group is more diverse than another, with no sound empirical support. None the less, it would seem likely that the evolution of some traits opened up opportunities for some groups to diversify disproportionately more than others. Thus, there is quantitative evidence that the adoption of phytophagy ('plant eating') has been associated with disproportionate diversification in insect groups (Mitter *et al.* 1988), whilst the adoption of a carnivorous parasitic lifestyle has not (Wiegmann *et al.* 1993).

2.3.3 Extinction

> *. . . strange fossil bones . . . have been found. . . . Everything seems to suggest that they represent vanished forms, animals that once existed and today no longer exist.* Count Buffon (1707–1788) (from Wendt 1970, p. 80)

The overall pattern of temporal change in biodiversity results from the difference between rates of speciation and rates of extinction. If species are being

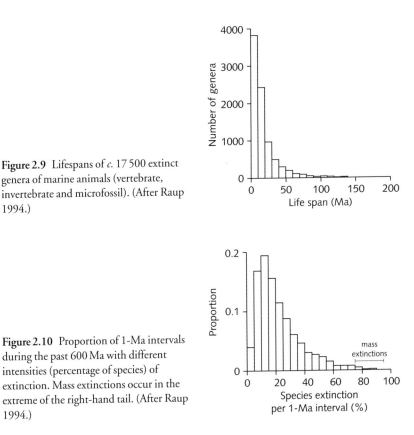

Figure 2.9 Lifespans of *c.* 17 500 extinct genera of marine animals (vertebrate, invertebrate and microfossil). (After Raup 1994.)

Figure 2.10 Proportion of 1-Ma intervals during the past 600 Ma with different intensities (percentage of species) of extinction. Mass extinctions occur in the extreme of the right-hand tail. (After Raup 1994.)

generated faster than they are becoming extinct, then the level of biodiversity will rise. When the level of extinction equals that of speciation an overall pattern of stability (stasis) will result. Hence if, or when, we observe stasis in the level of biodiversity this does not necessarily mean that nothing is happening; turnover in the identities of taxa through time could still be high. When the level of extinction exceeds speciation then biodiversity will decline, and if this persisted for a sufficient period then life would disappear from the Earth.

Over the history of life on Earth, in excess of 90% of all species are estimated to have become extinct. Raup (1994) found that the recorded lifespans of 17 500 genera of fossil marine animals were strongly right-skewed (Fig. 2.9). Most genera persisted for a relatively short time, whilst a few persisted for a very long period; the real pattern is probably even more skewed, as the very short-lived are unlikely even to be recorded in the fossil record. Compared with the duration of life on Earth, however, no genus survived for very long. The longest-lived persisted for about 160 Ma, or about 5% of the history of life.

Based on evidence from a variety of groups (both marine and terrestrial), the best present estimate is that the average species has had a lifespan (i.e. from the time a particular species appears in the fossil record until the time it disappears) of around 5–10 Ma (May *et al.* 1995). The intensity of extinction has varied markedly over time, with comparatively low levels during some periods and high ones during others (Fig. 2.10). The right-hand tail of this continuum comprises what have come to be known as the *mass extinctions* (the other periods comprise 'background' extinctions), albeit clearly they do not represent a distinct subset of periods. Although in these short intervals 75–95% of species alive at that time are estimated to have become extinct (Jablonski 1995), in sum the mass extinctions only account for about 4% of all extinctions in the last 600 Ma (Raup 1994). Their importance therefore lies not in their contribution to total extinctions, but in the disruptive effect they have had on the patterns of development of biodiversity. They reveal that marine and terrestrial biotas are not infinitely resilient and can be pushed beyond their limits by certain environmental stresses (Jablonski 1991). When levels of biodiversity recover, they often have a markedly different composition to those that preceded a mass extinction.

Although they are simply the tail of a continuum, it is highly unlikely that mass extinctions were simply the chance coincidence of extinctions of very large numbers of species. They had rather different causes. The end-Permian extinction, approximately 245 Ma ago, has variously been argued to have resulted from changes in sea level, climate change, oceanic anoxia and the explosion of a supernova. The end-Cretaceous extinction, best-known for the extinction of the ruling reptiles, was suggested to have been caused by the impact of an asteroid and the dust clouds which it generated.

Going Further

On your own
- Is it possible to attribute mass extinction events to particular causes?
- Read the following articles: Alvarez *et al.* (1980), Hallam (1987), Officer and Drake (1983) and Sharpton *et al.* (1992).
- Make a list of suggested causes for the major extinction at the end of the Cretaceous Period. Beside your list draw up two columns titled 'evidence for' and 'evidence against'.
- When you have finished, review the contents of both columns and see if you can answer the question posed above.

The history of biodiversity reveals that levels of biodiversity recover from mass-extinction events very rapidly (e.g. the families of marine invertebrates in Fig. 2.1) on an evolutionary time-scale, but the recovery and the re-establishment of some communities still typically requires 5–10 Ma (Jablonski 1995).

2.4 How many extant species are there?

If the diversity of life has increased through evolutionary time, how many species are presently extant? Although it has received substantial attention, the

Table 2.2 Approximate numbers of described species (in thousands) currently recognized and estimates of possible species richness for groups with more than 20 000 described species and/or estimated to include in excess of 100 000 species. The reliability of all estimates is likely to vary greatly. (After Hawksworth and Kalin-Arroyo 1995.)

	Described species	Number of estimated species			Accuracy of working figure
		High	Low	Working figure	
Viruses	4	1000	50	400	Very poor
Bacteria	4	3000	50	1000	Very poor
Fungi	72	2700	200	1500	Moderate
'Protozoa'	40	200	60	200	Very poor
'Algae'	40	1000	150	400	Very poor
Plants	270	500	300	320	Good
Nematodes	25	1000	100	400	Poor
Arthropodsi					
Crustaceans	40	200	75	150	Moderate
Arachnids	75	1000	300	750	Moderate
Insects	950	100000	2000	8000	Moderate
Molluscs	70	200	100	200	Moderate
Chordates	45	55	50	50	Good
[Others	115	800	200	250	Moderate]
Totals	1750	111655	3635	13620	Very poor

importance of this question perhaps has less to do with the usefulness of the actual answer than with the challenge it poses to our understanding of how biodiversity is distributed amongst different groups of organisms and across the Earth.

On the face of it, the best way of finding out how many extant species there are would be simply to count them! Given the enormity of such a task, all of the attempts at answering this question have employed indirect measures and, in the process, have made major assumptions of one kind and another (for reviews see May 1988, 1990, 1994; Hammond 1995; Pimm *et al.* 1995a; Stork 1997). The best working estimate of extant species numbers is one of around 13.5 million, with upper and lower estimated numbers of about 3.5 and 111.5 million species, respectively (Table 2.2) (Hawksworth & Kalin-Arroyo 1995; see also World Conservation Monitoring Centre 1992; Hammond 1995). The upper boundary appears wildly improbable, if for no other reason than that it is not obvious where all the 'missing' species are to be found! Evidence in support of the working estimate is becoming increasingly convincing, albeit categorical demonstrations of its validity do not exist.

Going Further

On your own
• Read the following references: May (1988, 1990, 1992a) and Pimm *et al.* (1995a).
• Write in note form a description of how we can best calculate how many species there are.

The more rigorous estimates of global species numbers are arrived at, implicitly or explicitly, largely by summation of estimates from a variety of sources (e.g. canvassing experts in particular taxonomic groups, extracting relevant data from published scientific papers or monographs). The major uncertainties remain in figures for particular taxonomic groups (e.g. viruses, bacteria, nematodes, mites), functional groups (e.g. parasites) and habitats or biomes (e.g. soils, deep ocean benthos). Indeed, the relative contributions of some groups and areas compared with others continues to be debated, sometimes vigorously (e.g. see Hammond 1995).

What is clear, however, is that if we construct a breakdown of the estimated number of species in various major groups of organisms, to a first approximation every animal is an insect (see Table 2.2)!

Going Further

On your own or as part of a group
• Read Grassle (1989) and May (1992b, 1994).
• Summarize and critically evaluate the evidence for diversity in the deep sea being as great as found anywhere else on Earth.
• (optional) Prepare a poster presentation that in a concise and engaging way presents an objective assessment of our current understanding of deep-sea diversity.

In terms of constructing an inventory of different individual species we are severely hampered by the fact that only a fraction of the total number of species has been formally taxonomically described. Even determining how large a fraction is complicated by the absence of a definitive listing of all described species and their status (e.g. whether they are presently regarded as valid or not). Best estimates are that approximately 1.75 million living species have been described, i.e. about 13% of the estimated total number of extant species (Table 2.2, Hawksworth & Kalin-Arroyo 1995).

Additional species are being described at a rate of about 13 000 per annum; the breakdown for major groups of animals is given in Fig. 2.11. This rate has remained remarkably constant over the past decade or so (Hawksworth &

Box 2.2 New mammals

It is often thought that the age of description is finished and that the finding of a new species of mammal is something of a novelty. While it does seem that the major period of discovery has passed, the number of new mammal species still being described is not insignificant (Fig. 1).

Sixteen new, living species of large mammals alone have been discovered during the period 1937 to the present, about three per decade (Pine 1994): two porpoises (*Lagenodelphis hosei, Phocoena sinus*), four beaked whales (*Tasmacetus shepherdi, Mesoplodon ginkgodens, M. carlhubbsi, M. peruvianus*), a wild pig (*Sus heureni*), a peccary (*Catagonus wagneri*), four deer (*Mazama chunyi, Moschus fuscus, Muntiacus atherodes, Muntiacus gongshanensis*), the kouprey (*Bos sauveli*), a gazelle (*Gazella bilkis*), a wild sheep (*Pseudois schaeferi*) and a 'bovid' (*Pseudoryx nghetinhensis*). Not only is finding a new mammal important for our understanding of mammal diversity and biogeography, and for its conservation, but often associated with the animal are a host of unknown parasites and pathogens as well. However, given the present rate of loss of biodiversity it is not inconceivable that many mammal species will become extinct even before they are described.

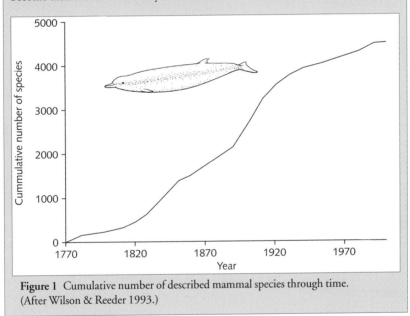

Figure 1 Cumulative number of described mammal species through time. (After Wilson & Reeder 1993.)

Kalin-Arroyo 1995). However, the particular taxonomic group a scientist chooses to work on is largely a matter of serendipity and personal choice. Consequently our catalogue of biodiversity has grown in a somewhat haphazard fashion. Equally, the species that have been described are far from a random subset of all species. On average they are larger-bodied, more abundant (locally or regionally), more widely distributed, occupy a larger number of habitats or

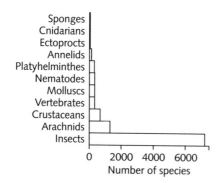

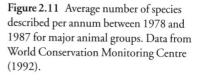

Figure 2.11 Average number of species described per annum between 1978 and 1987 for major animal groups. Data from World Conservation Monitoring Centre (1992).

life zones and derive disproportionately from temperate zones. And even where species have been formally described this should not be taken to mean that we know much about them. For example, a recent estimate suggests that about 40% of described species of beetles are each known from only a single locality (May *et al.* 1995).

The gap between the number of described animal species and the estimated total number of extant species is due predominantly to our ignorance of small-bodied invertebrate taxa. However, it should not be forgotten that in some areas even some vertebrate taxa are still comparatively poorly known (e.g. <50% of African freshwater fishes are estimated to have been described; Ribbink 1994) and that new species in many vertebrate taxa continue to be discovered (Box 2.2).

There is no likelihood that in the foreseeable future the disparity between the total number of extant species and the number of species that have been described will be markedly closed. This is simply because the taxonomic work-force does not exist to perform the task. In fact, the present work-force is actually in decline (Gaston & May 1992). In the face of this lamentable state of affairs, fulfilment of the task of describing all species will remain a far-distant prospect. What is required is a planned targeting of key groupings, taxa and geographical areas for taxonomic description designed to give a better understanding of the important questions in the study of biodiversity and other fields, rather than the *ad hoc* accumulation of taxa we see at present.

2.5 Recent and future extinctions

2.5.1 Species losses

As well as seeking to catalogue and measure it, humankind has for a long period been instrumental in the erosion of biodiversity. Arguably, early humans were directly responsible for the extinction of many large mammals during the late Pleistocene, and indirectly for the loss of many smaller ones

Figure 2.12 The relationship between the percentage of the recent avifauna of Pacific islands that is extinct or currently endangered and the duration of human occupancy of those islands. The Marianas have an unusually high number of modern losses, as a result of recent colonization by the brown tree snake. (After Pimm *et al.* 1995b.)

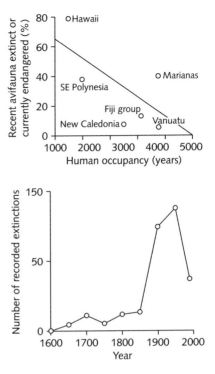

Figure 2.13 The number of recorded global extinctions of animal species since *c.* 1600 for which a date is known. (After Smith *et al.* 1993a.)

(Martin 1984; Owen-Smith 1987). Large numbers of avian extinctions followed the colonization of tropical Pacific islands by prehistoric peoples, with roughly half the landbird species on each island group estimated to have been exterminated (Pimm *et al.* 1995b; Steadman 1995). Indeed, the proportion of the avifauna on these island groups that has recently become extinct or is now endangered or in immediate danger of extinction is less where human occupancy has been longest. This suggests that those areas colonized first have already lost most of the species that are sensitive to human activities (Fig. 2.12).

Our knowledge of human impact on biodiversity inevitably improves, although it remains sketchy, as we move closer to the present. Since 1600 there have been over 1000 recorded extinctions of animal and plant species. Roughly half of these extinctions took place this century. Smith *et al.* (1993a) document a significant rise in the rate of recorded species extinctions for well-known groups of animals over the past 400 years, with a sharp increase in the nineteenth century coinciding with European colonial expansion (Fig. 2.13). A global decline in the rate since about 1950 may in part reflect the growth of conservation activities, but may equally (and perhaps more likely) be due to the introduction of more stringent criteria for deciding that a species is genuinely extinct (rather than that it has simply gone unrecorded).

Interesting as the data on recorded extinctions may be, they undoubtedly underestimate the true levels of species extinctions. Amongst animals, available information is strongly biased towards birds and mammals and away from invertebrates (as is our broad understanding of biodiversity; see Section 2.4). More generally, available information is geographically biased towards islands and developed nations. Undoubtedly, although impossible to demonstrate categorically, many undescribed species have become extinct without us even being aware of their existence, and the loss of many species that have been described has doubtless gone unrecorded.

As we have seen (Section 2.3.3), the average lifespan of any species in the fossil record is estimated to be around 5–10 Ma. For birds and mammals, rates of documented extinction over the past century correspond to species' lifespans of around 10 000 years (May *et al.* 1995). Although the calculations are inevitably very rough and ready, projection of impending extinctions, if current trends continue, suggest a lifespan for bird and mammal species of 200–400 years! These figures may perhaps be regarded as representative of a broad range of organisms, in which case impending extinction rates are at least four orders of magnitude faster than the background rates seen in the fossil record.

Going Further

On your own or as part of a group
• As a professional biologist you are to produce a discussion document on the best way to estimate future extinction rates of species. The target audience is political decision-makers, so the document must be short, concise and written in non-technical language.

References
Myers (1990), Smith *et al.* (1993b), Mace (1995), May *et al.* (1995), Pimm *et al.* (1995b)

Merely recording numbers of extinctions may underestimate the effects of past human activity on biodiversity, through a process known as extinction debt. That is, a time-lag between these activities, such as deforestation, and their full effects being realised. Individuals of large-bodied species, for example, may persist after the populations to which they belong have ceased to be viable, because they are long-lived. The species is effectively extinct; it just doesn't know it yet! Brooks and Balmford (1996) document an example of extinction debt in the Atlantic forests of South America. Here, whilst nearly 90% of the forest has been cleared, no bird species has so far been shown to have become extinct as a result, contrary to the predictions of species–area relationships (see Section 3.2.1). However, the number of species presently recognized as being highly threatened with extinction is similar to that predicted to become extinct from deforestation. It would seem that without immediate conservation action these species will inevitably soon be lost.

Table 2.3 Summary of the numbers of animal species in major taxonomic groups that are listed as extinct, extinct in the wild (the species has been extirpated from its natural habitat but survives in captivity or as a naturalized population outside its past range) or globally threatened with extinction, according to the 1996 IUCN Red List. (After IUCN 1996.)

	Extinct	Extinct in the wild	Threatened
Mammalia	86	3	1096
Aves	104	4	1107
Reptilia	20	1	253
Amphibia	5	0	124
Cephalaspidomorphi	1	0	3
Elasmobranchii	0	0	15
Actinopterygii	80	11	715
Sarcopterygii	0	0	1
Arachnida	0	0	10
Chilopoda	0	0	1
Crustacea	9	1	407
Insecta	72	1	537
Onychophora	3	0	6
Oligochaeta	0	0	5
Polychaeta	0	0	1
Bivalvia	12	0	114
Gastropoda	216	9	806
Enopla	0	0	2
Turbellaria	1	0	0
Anthozoa	0	0	2

One of the bases for estimating impending extinctions is information on numbers of species that have been listed as being threatened with global extinction in the near future. The most recent figures for animals are given in Table 2.3. Again, and by now perhaps not surprisingly, these figures are highly biased, in much the same ways as are those of recorded extinctions. Only for birds and mammals has the threat status of virtually all species been evaluated.

2.5.2 Population losses and declines

The extinction of a species equates to the loss of all local populations, and typically follows a marked decline in overall abundance. Without necessarily being associated with species extinction, the extinction of individual local populations and declines in species' local abundances both represent potentially insidious forms of erosion of biodiversity (Ehrlich & Daily 1993; Ehrlich 1995). Population losses, in particular, will tend to reduce the taxonomic, genetic and functional diversity of sites (see Table 1.1), and perhaps the performance of ecosystems (see Section 4.4.2).

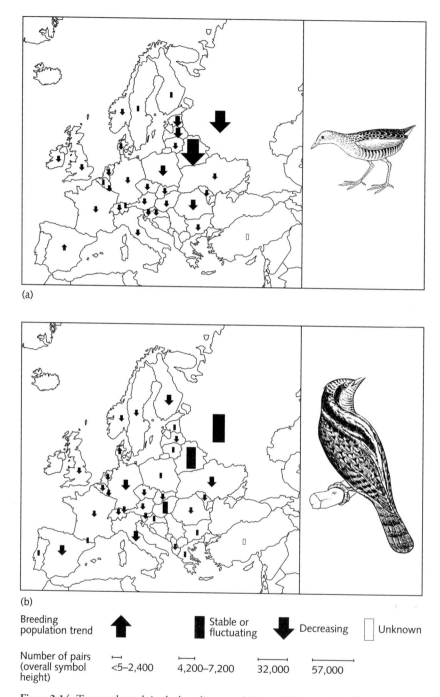

Figure 2.14 Temporal trends in the breeding populations of (a) corncrake *Crex crex* and (b) wryneck *Jynx torquilla* in countries of Europe. (After Tucker & Heath 1994.)

Population losses and declines are undoubtedly very high in some regions and groups (Fig. 2.14). For example, 38% (195 species) of the avifauna of Europe is regarded as having an unfavourable conservation status (in descending order of threat: endangered, vulnerable, rare, declining, localized or insufficiently known). In the majority of cases this is because of substantial declines in the European populations of species, believed mostly to be linked to changes in land-use and land management (Tucker & Heath 1994).

2.5.3 Biodiversity in crisis

The present and projected losses of biodiversity (as reflected in species and population numbers and other measures) have been termed a 'crisis' (Wilson 1985). One definition of a crisis is 'a state of affairs in which a decisive change for better or worse is imminent'. There seems little reason for believing that this is not the case *vis-à-vis* biodiversity, and absolutely no reason to believe that the change is for better rather than worse. Indeed, one must wonder to what extent the decisive change is current rather than imminent. Serious loss of biodiversity is presently taking place.

2.6 Ups as well as downs?

There is a harsh twist to the biodiversity crisis. In addition to driving species extinct, human actions have also served, intentionally or accidentally, to introduce species to areas in which they would not naturally have occurred. This is especially true of areas with long histories of human interchange. Ebenhard (1988) observed that, just on the basis of reports in the literature, worldwide humans have introduced 330 species of birds and mammals alone, in 1559 separate cases. The effects of such introductions, however, have not been to offset the extinctions that have resulted from human activities. At a regional scale introductions may tend to increase biodiversity (adding to the pre-existing biota), but they also tend to homogenize biodiversity between regions (increasing the similarity of their biota, a process which by definition is also engendered by the extinction of rarer species). For example, several species now widespread across the Earth, such as the feral cat *Felis catus*, common rat *Rattus norvegicus* and house mouse *Mus domesticus*, have become so through human agency. At a more local scale, reductions in biodiversity may (though far from always) result, as introduced species outcompete (directly or indirectly) or prey on the pre-existing biota. One of the most dramatic examples concerns the brown tree snake *Bioga irregularis*, which since its introduction has caused a precipitous decline in the avifauna of the island of Guam. Of a resident avifauna of 18 native species and a further seven species that were themselves introduced, seven species are probably now extinct and four others are so rare that their survival seems unlikely (Savidge 1987; Pimm 1991).

2.7 Summary

1 There has been an increase in biodiversity, from the appearance of the first organism up to the present day, despite the fact that more than 90% of all species that have existed have become extinct.
2 This increase has not been continuous but is composed of radiations and stabilizations, punctuated by mass extinctions, of different groups at different times.
3 Present-day biodiversity stands at an estimated 13.5 million species with only 1.75 million of these currently described.
4 This present extent of biodiversity is contributed by only a relatively few groups of organisms. Conversely, most groups are not particularly diverse.
5 Humankind has been, and continues to be, instrumental in the erosion of biodiversity. Current extinction rates are possibly four orders of magnitude greater than the background rates seen in the fossil record.
6 The biodiversity crisis may well be current rather than imminent.
7 Population losses and declines, which are high in some regions and for some groups of organisms, represent a potentially insidious form of erosion of biodiversity.
8 Human interchange has been responsible for introducing species to new areas, often reducing local biodiversity and homogenizing biodiversity between regions.

2.8 Making connections

Like the history of humankind, the history of biodiversity is not a simple one. However, some of the major signals are relatively clear. In the chapter that follows, we leave temporal patterns and go on to examine spatial patterns detectable in present biodiversity (i.e. where organisms now occur). As already indicated (Section 2.5), some of the major forces shaping current spatial patterns are the actions and influence of humankind, but we will largely defer our discussion of anthropogenic effects until the final chapter (Chapter 5) when we consider how best to maintain biodiversity.

2.9 References

Alvarez, L.W., Alvarez, W., Asaro, F. & Michel, H.V. (1980) Extraterrestrial causes for the Cretaceous–Tertiary extinctions. *Science* **208**, 1095–1108.
Ayer, A.J. (1973) *The Central Questions of Philosophy*. Weidenfeld & Nicholson, London.
Benson, M.J. (1985) Mass extinction among non-marine tetrapods. *Nature* **316**, 811–814.
Brooks, T. & Balmford, A. (1996) Atlantic forest extinctions. *Nature* **380**, 115.
Conway-Morris, S. (1979) The Burgess Shale (middle Cambrian) fauna. *Annual Review of Ecology and Systematics* **10**, 327–349.
Darwin, C. (1859) *The Origin of Species by Means of Natural Selection*. John Murray, London.

Dial, K.P. & Marzluff, J.M. (1989) Nonrandom diversification within taxonomic assemblages. *Systematic Zoology* **38**, 26–37.

Dodson, P. (1990) Counting dinosaurs: how many kinds were there? *Proceedings of the National Academy of Sciences USA* **87**, 7608–7612.

Ebenhard, T. (1988) Introduced birds and mammals and their ecological effects. *Swedish Wildlife Research* **13** (4), 1–107.

Ehrlich, P.R. (1995) The scale of the human enterprise and biodiversity loss. In: *Extinction Rates* (eds J.H. Lawton & R.M. May), pp. 214–226. Oxford University Press, Oxford.

Ehrlich, P.R. & Daily, G.C. (1993) Population extinction and saving biodiversity. *Ambio* **22**, 64–68.

Funch, P. & Kristensen, R.H. (1995) Cycliophora is a new phylum with affinities to Entoprocta and Ectoprocta. *Nature* **378**, 711–714.

Gaston, K.J. & May, R.M. (1992) The taxonomy of taxonomists. *Nature* **356**, 281–282.

Gould, S.J. (1989) *Wonderful Life. The Burgess Shale and the Nature of History*. Norton, New York.

Grassle, J.F. (1989) Species diversity in deep-sea communities. *Trends in Ecology and Evolution* **4**, 12–15.

Hallam, A. (1987) End-Cretaceous mass extinction event: argument for terrestrial causation. *Science* **238**, 1237–1242.

Hammond, P.M. (1995) Described and estimated species numbers: an objective assessment of current knowledge. In: *Microbial Diversity and Ecosystem Function* (eds D. Allsopp, D.L. Hawksworth & R.R. Colwell), pp. 29–71. CAB International, Wallingford.

Hawksworth, D.L. & Kalin-Arroyo, M.T. (1995) Magnitude and distribution of biodiversity. In: *Global Biodiversity Assessment* (ed. V.H. Heywood), pp. 107–191. Cambridge University Press, Cambridge.

IUCN (1996) *1996 IUCN Red List of Threatened Animals*. IUCN, Gland, Switzerland.

Jablonski, D. (1991) Extinctions: a paleontological perspective. *Science* **253**, 754–757.

Jablonski, D. (1995) Extinctions in the fossil record. In: *Extinction Rates* (eds J.H. Lawton & R.M. May), pp. 25–44. Oxford University Press, Oxford.

Labandeira, C.C. & Sepkoski Jr, J.J. (1993) Insect diversity in the fossil record. *Science* **261**, 310–315.

Laverack, M.S. & Dando, J. (1987) *Lecture Notes on Invertebrate Zoology*, 3rd edn. Blackwell Scientific Publications, Oxford.

Mace, G.M. (1995) Classification of threatened species and its role in conservation planning. In: *Extinction Rates* (eds J.H. Lawton & R.M. May), pp. 197–213. Oxford University Press, Oxford.

Margulis, L., Schwartz, K.V. & Dolan, M. (1994) *The Illustrated Five Kingdoms. A Guide to the Diversity of Life on Earth*. Harper Collins, New York.

Martin, P.S. (1984) Prehistoric overkill: the global model. In: *Quaternary Extinctions: a Prehistoric Revolution* (eds P.S. Martin & R.G. Klein), pp. 354–403. University of Arizona Press, Tucson.

May, R.M. (1988) How many species are there on Earth? *Science* **241**, 1441–1449.

May, R.M. (1990) How many species? *Philosophical Transactions of the Royal Society of London Series B* **330**, 293–304.

May, R.M. (1992a) How many species inhabit the Earth? *Scientific American* **267**, 42–48.

May, R.M. (1992b) Bottoms up for the oceans. *Nature* **357**, 278–279.

May, R.M. (1994) Conceptual aspects of the quantification of the extent of biological diversity. *Philosophical Transactions of the Royal Society of London Series B* **345**, 13–20.

May, R.M., Lawton, J.H. & Stork, N.E. (1995) Assessing extinction rates. In: *Extinction Rates* (eds J.H. Lawton & R.M. May), pp. 1–24. Oxford University Press, Oxford.

Mitter, C., Farrell, B. & Wiegmann, B. (1988) The phylogenetic study of adaptive zones: has phytophagy promoted insect diversification? *American Naturalist* **132**, 107–128.

Myers, N. (1990) Mass extinctions: what can the past tell us about the present and the future? *Palaeogeography, Palaeoclimatology, Palaeoecology (Global and Planetary Change Section)* **82**, 175–185.

Niklas, K.J. (1986) Large-scale changes in animal and plant terrestrial communities. In: *Patterns and Processes in the History of Life* (eds D.M. Raup & D. Jablonski), pp. 383–405. Springer-Verlag, Berlin.

Officer, C.B. & Drake, C.L. (1983) The Cretaceous–Tertiary transition. *Science* **219**, 1383–1390.

Owen-Smith, N. (1987) Pleistocene extinctions: the pivotal role of megaherbivores. *Paleobiology* **13**, 351–362.

Pimm, S.L. (1991) *The Balance of Nature? Ecological Issues in the Conservation of Species and Communities*. Chicago University Press, Chicago.

Pimm, S.L., Russell, G.J., Gittleman, J.L. & Brooks, T.M. (1995a) The future of biodiversity. *Science* **269**, 347–350.

Pimm, S.L., Moulton, M.P. & Justice, L.J. (1995b) Bird extinctions in the central Pacific. In: *Extinction Rates* (eds J.H. Lawton & R.M. May), pp. 75–87. Oxford University Press, Oxford.

Pine, R.H. (1994) New mammals not so seldom. *Nature* **368**, 593.

Raup, D.M. (1994) The role of extinction in evolution. *Proceedings of the National Academy of Sciences USA* **91**, 6758–6763.

Ribbink, A.J. (1994) Biodiversity and speciation of freshwater fishes with particular reference to African cichlids. In: *Aquatic Ecology: Scale, Pattern and Process* (eds P.S. Giller, A.G. Hildrew & D.G. Raffaelli), pp. 261–288. Blackwell Science, Oxford.

Savidge, J.A. (1987) Extinction of an island avifauna by an introduced snake. *Ecology* **68**, 660–668.

Schopf, J.W. (ed.) (1992) *Major Events in the History of Life*. Jones & Bartlett, Boston.

Sepkoski Jr, J.J. (1992) Phylogenetic and ecologic patterns in the Phanerozoic history of marine biodiversity. In: *Systematics, Ecology, and the Biodiversity Crisis* (ed. N. Eldredge), pp. 77–100. Columbia University Press, New York.

Sharpton, V.L., Dalrymple, G.B., Marin, L.E., Ryder, G., Schuraytz, B.C. & Urrutia-Fucugauchi, J. (1992) New links between Chicxulub impact structure and the Cretaceous/Tertiary boundary. *Nature* **359**, 819–821.

Signor, P.W. (1990) The geologic history of diversity. *Annual Review of Ecology and Systematics* **21**, 509–539.

Slowinski, J.B. & Guyer, C. (1989) Testing the stochasticity of patterns of organismal diversity: an improved null model. *American Naturalist* **134**, 907–921.

Smith, F.D.M., May, R.M., Pellew, R., Johnson, T.H. & Walter, K.R. (1993a) How much do we know about the current extinction rate? *Trends in Ecology and Evolution* **8**, 375–378.

Smith, F.D.M., May, R.M., Pellew, R., Johnson, T.H. & Walter, K.S. (1993b) Estimating extinction rates. *Nature* **364**, 494–496.

Steadman, D.W. (1995) Prehistoric extinctions of Pacific island birds: biodiversity meets zooarchaeology. *Science* **267**, 1123–1131.

Stork, N.E. (1997) Measuring global biodiversity and its decline. In: *Biodiversity II* (eds M.L. Reaka-Kudla, D.E. Wilson & E.O. Wilson), pp. 41–68. Joseph Henry Press, Washington, DC.

Tappan, H. & Loeblich Jr, A.R. (1973) Evolution of the oceanic plankton. *Earth Science Reviews* 9, 207–240.

Tucker, G.M. & Heath, M.F. (1994) *Birds in Europe: their Conservation Status.* BirdLife International, Cambridge.

Van Valkenburgh, B. & Janis, C.M. (1993) Historical diversity patterns in North American large herbivores and carnivores. In: *Species Diversity in Ecological Communities: Historical and Geographical Perspectives* (eds R.E. Ricklefs & D. Schluter), pp. 330–340. Chicago University Press, Chicago.

Wendt, H. (1970) *Before the Deluge.* Paladin, London.

Whittaker, R.H. & Margulis, L. (1978) Protist classification and the Kingdoms of organisms. *Biosystems* 10, 3–18.

Wiegmann, B.M., Mitter, C. & Farrell, B. (1993) Diversification of carnivorous parasitic insects: extraordinary radiation or specialized dead end? *American Naturalist* 142, 737–754.

Williams, D.M. & Embley, T.M. (1996) Microbial diversity: domains and kingdoms. *Annual Review of Ecology and Systematics* 27, 569–595.

Wilson, D.E. & Reeder, D.M. (eds) (1993) *Mammal Species of the World: a Taxonomic and Geographic Reference.* Smithsonian Institution Press, Washington, DC.

Wilson, E.O. (1985) The biological diversity crisis: a challenge to science. *BioScience* 35, 700–706.

Woese, C.R., Kandler, O. & Wheelis, M.L. (1990) Towards a natural system of organisms: proposal for the domains Archaea, bacteria and Eucarya. *Proceedings of the National Academy of Sciences USA* 87, 4576–4579.

World Conservation Monitoring Centre (1992) *Global Biodiversity: Status of the Earth's Living Resources.* Chapman & Hall, London.

2.10 Further reading

Archibald, J.D. (1996) *Dinosaur Extinction and the End of an Era: What the Fossils Say.* Columbia University Press, New York. (*A readable book, especially for the non-specialist, on the extinction of dinosaurs and other vertebrates.*)

Barnes, R.D. (1989) Diversity of organisms: how much do we know? *American Zoologist* 29, 1075–1084. (*A bit dated now, but still worth reading.*)

Brusca, R.C. & Brusca, G.J. (1990) *Invertebrates.* Sinauer Associates, Sunderland, Massachussetts. (*A good, general textbook on invertebrate biology.*)

Crawley, M.J. (1997) Biodiversity. In: *Plant Ecology*, 2nd edn (ed. M.J. Crawley), pp. 595–632. Blackwell Science, Oxford. (*A very useful overview of the biodiversity of plants, with sections on temporal dynamics, and aliens.*)

Gee, H. (1996) *Before the Backbone: Views on the Origin of the Vertebrates.* Chapman & Hall, London. (*Extremely well written, state-of-the-art discussion of vertebrate origins.*)

Hammond, P.M. (1994) Practical approaches to the estimation of the extent of biodiversity in speciose groups. *Philosophical Transactions of the Royal Society of London Series B* 345, 119–136. (*A particular viewpoint on estimating the species richness of the very largest groups.*)

Lawton, J.H. & May, R.M. (eds) (1995) *Extinction Rates.* Oxford University Press, Oxford. (*The definitive book on extinction.*)

Minelli, A. (1993) *Biological Systematics: the State of the Art.* Chapman & Hall, London. (*An accessible overview of biological systematics, including the state of taxonomy of major groups.*)

Nee, S., Barraclough, T.G. & Harvey, P.H. (1996) Temporal changes in biodiversity:

detecting patterns and identifying causes. In: *Biodiversity: a Biology of Numbers and Difference* (ed. K.J. Gaston), pp. 230–252. Blackwell Science, Oxford. (*A fast-developing area exploring the dynamics of biodiversity based on phylogeny, something that we don't really deal with here.*)

Niklas, K.J. & Tiffney, B.H. (1994) The quantification of plant biodiversity through time. *Philosophical Transactions of the Royal Society of London Series B* **345**, 35–44. (*How do you actually quantify the temporal diversity of plants?*)

Quicke, D.L.J. (1993) *Principles and Techniques of Contemporary Taxonomy.* Blackie, London. (*Putting a name to a species – a modern treatment of principles and practice.*)

Rosenzweig, M.L. (1995) *Species Diversity in Space and Time.* Cambridge University Press, Cambridge. (*A major review of some of the principal patterns, and the mechanisms which underpin them.*)

Ruppert, E.E. & Barnes, R.D. (1991) *Invertebrate Zoology,* 6th edn. Saunders College Publishing, Fort Worth, Texas. (*Still, to our mind,* the *invertebrate textbook.*)

Schopf, J.W. (ed.) (1992) *Major Events in the History of Life.* Jones & Bartlett, Boston. (*Contains some very good chapters, written by experts but mainly pitched at undergraduate level, introducing the oldest fossils (Chapter 2), the evolution of the earliest animals (Chapter 3) and diversification of the vertebrates (Chapter 5).*)

Williamson, M. (1996) *Biological Invasions.* Chapman & Hall, London. (*Makes sense of a large and bewildering literature.*)

Young, J.Z. (1981) *The Life of Vertebrates,* 3rd edn. Oxford University Press, Oxford. (*The classic introduction to the biology of vertebrates, including their diversity.*)

3 Mapping biodiversity

3.1 Introduction

It has long been evident to naturalists that the ordinary political divisions of the earth's surface do not correspond with those based on the geographical distribution of animal life. Europe, for instance, the most important of all the continents politically speaking, is for zoological geographers, as well as for physical, but a small fragment of Asia. . . . Let us then therefore dismiss from our minds for the moment the ordinary notions of both physical and political geography, and consider how the earth's surface may be most naturally divided into Primary Regions, taking the amount of similarity and dissimilarity of animal life as our sole guide. Sclater & Sclater (1899)

If you want to see a lava heron *Butorides striatus sundevalli* in the wild you will have to organize a trip to the Galapagos Islands where it is endemic; in fact, this particular subspecies very rarely leaves the lava rocks found on the shores of these islands. Likewise, if you want to observe terrestrial crabs in their native environment then you must travel to the tropics (assuming that you are not already sitting in the equatorial sunshine reading this book) as these animals are not really found in cold temperate zones. Similarly, if you want to find as many different insect species as possible, then exploring a moist tropical forest is infinitely preferable to a cold temperate one. All this, of course, seems self-evident. But it nicely illustrates a point: biodiversity is not distributed evenly across the Earth, beneath it, or through the media (e.g. air, water) which blanket it. Rather, there is a richly textured surface of 'hotspots' of high biodiversity, 'coldspots' of low diversity, and many 'warmish spots' in between.

Despite the increased attention given to biodiversity in the last few years both by the press and in the form of scientific research, we remain far from being able to produce a complete set of maps that document the detailed occurrence of life on Earth. That is, we do not yet have an atlas of biodiversity. This should not be a surprise. For no geographical area, even those of only moderate size, do we even have a complete count of all the species (across all taxa) that occur there. None the less, many patterns of spatial variation in biodiversity have been documented. Although these patterns are interesting in their own right, they may also provide the 'skeleton' on which a biodiversity atlas may one day be built.

In this chapter, we do five things. First, we address some issues regarding the effects of spatial scale on observed levels of biodiversity. Second, we identify spatial patterns in extremes of high and low biodiversity. Third, we identify spatial gradients in biodiversity. Fourth, we consider relationships between biodiversity and environmental variables. Finally, we discuss spatial congruence in the biodiversity of different groups, and the prospects for determining 'the big picture' which will allow us to make generalizations about the distribution of life on Earth.

Throughout, examples will again be drawn predominantly from a few taxonomic groups of animals and plants. This is more a matter of necessity than of convenience. Very little empirical information is available about spatial patterns in the biodiversity of most of the highly speciose groups, such as the bacteria, fungi and insects (a point we will return to at the close of the chapter). As explained in Chapter 1, we again use species richness as our main measure, and common currency, of biodiversity. This is not a forced arrangement but is essential, because understanding of most of the patterns discussed is founded on species richness. Where some of the patterns in biodiversity have been examined using other measures (e.g. numbers of higher taxa) similar results to those found for species richness have in the main been obtained. We will also draw on a few examples of these studies.

3.2 Issues of scale

3.2.1 Species–area relationships

The principal pattern of spatial scaling of biodiversity is the species–area relationship. As the size of a geographical area increases, so does the number of species it contains (MacArthur & Wilson 1967; Williamson 1988; Rosenzweig 1995); the numbers of higher taxa usually follow a similar pattern. For example, across Europe, the number of species of lumbricid earthworms increases from less than 10 in areas of about $100 \, m^2$ to more than 50 in areas of $500\,000 \, km^2$ or more (Fig. 3.1a). Likewise, the number of plant species on Australian islands increases with the size of those islands (Fig. 3.1b).

Over the range of sizes of areas encompassed by small pieces of uniform habitat to continents and beyond, the processes generating the increase in diversity with area vary markedly. Indeed, Rosenzweig (1995) argues that the species–area relationship is actually not one pattern but four.

1 Species–area relationships among tiny pieces of a single biota: these may result from the sampling of more individual organisms, but below some threshold area biodiversity may not always be a simple function of area (see Fig. 3.1a,b).

2 Species–area relationships among large pieces of a single biota: these result principally from larger areas containing more habitats.

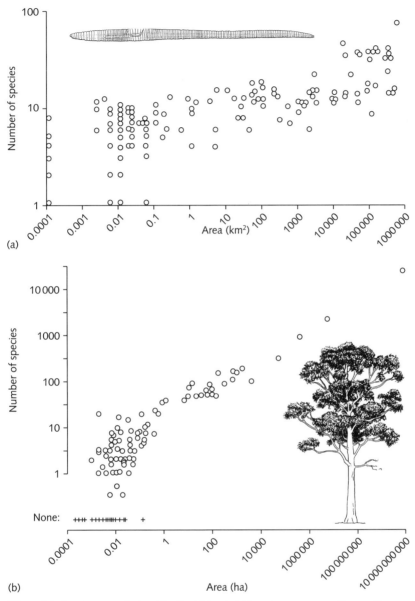

Figure 3.1 Species–area relationships for (a) lumbricid earthworms across Europe (from data in Judas 1988) and (b) plants on Australian islands. In both cases axes are logarithmically transformed. (After Rosenzweig 1995.)

3 Species–area relationships among islands of one archipelago: these result from larger areas containing more habitats, and from the interplay between the immigration and extinction of species on islands.
4 Species–area relationships among provinces: these result from a greater rate of speciation and a lower rate of extinction in larger areas.

Conventionally, in exploring relationships between species richness (the dependent variable) and area (the independent variable) both are logarithmically transformed, which serves approximately to linearize the interaction between the two. Presented thus, the different patterns above tend, on average, to have rather different slopes, with those for provinces being the steepest, followed by those for large pieces of a single biota, followed by those for islands.

Species–area relationships are of practical significance with regard to biodiversity, because they predict that as the area of a habitat is reduced it will tend to lose species. Combined with the major impact human activity is having on the extent of different habitats (see Section 5.2), it is thus not surprising that habitat loss constitutes one of the major determinants of present-day extinction rates (see Sections 2.5.1 & 2.5.2).

Many of the changes that humans are making to the landscape involve not simply the reduction of the areas of some habitats and the expansion of others, but also the fragmentation of the former into pieces. This results in a landscape consisting of (often small) remnant areas of native vegetation embedded in a matrix of agricultural and developed land. Fragmentation results in change in the physical environment within patches (e.g. in fluxes of radiation, water and nutrients) and in biogeographic changes (e.g. in isolation and connectivity), which have important consequences for the biota (Saunders *et al.* 1991).

Differences in the sizes of areas have, with some important exceptions (e.g. see Section 3.3.1), a pervasive influence on most spatial patterns in biodiversity. This must be borne in mind in much of the subsequent discussion in this chapter. However, the species–area relationship may sometimes be obscured or even reversed by some of the other patterns we will discuss, especially that between biodiversity and latitude (Section 3.4.1). For example, Costa Rica (in Central America) is estimated to have 1500–2000 species of butterfly whilst Britain has only about 60, and yet Costa Rica has less than one-sixth the land area of Britain.

3.2.2 Local–regional diversity relationships

Although it is true that smaller areas tend to contain fewer species than larger areas (see Section 3.2.1), the species richness of a small area is not independent of that of the larger area in which it is embedded. Local species richness tends to be an increasing function of regional richness, such that local areas in very diverse regions tend to have greater levels of diversity than do local areas in regions of low diversity (Cornell & Lawton 1992). In other words, just as an entire desert will contain fewer species of insect than an entire forest, so a small area of the desert will contain fewer species of insect than will a similar-sized small area of the forest. As a less sweeping example, the number of bird species at sites in the Caribbean is an increasing function of the total number of bird species in the regions in which those sites occur (Fig. 3.2). Similarly, the local

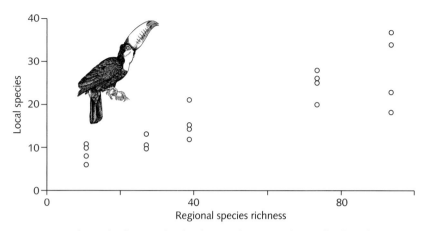

Figure 3.2 Relationship between local and regional species richness of birds in the Caribbean. (After Ricklefs 1987.)

species richness of lacustrine fish in North America is an increasing function of their regional richness (Griffiths 1997).

The relationship between local and regional biodiversity underpins the crucial observation that temporal changes in global biodiversity tend to be reflected in local biodiversity, and vice versa (see Section 2.3.1).

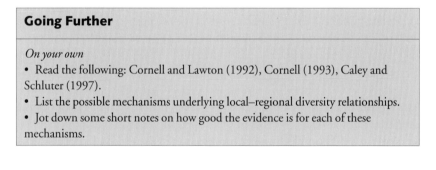

Going Further

On your own
• Read the following: Cornell and Lawton (1992), Cornell (1993), Caley and Schluter (1997).
• List the possible mechanisms underlying local–regional diversity relationships.
• Jot down some short notes on how good the evidence is for each of these mechanisms.

3.3 Extremes of high and low diversity

In asking about spatial patterns of biodiversity, a convenient starting point is to identify where the extreme values tend to lie.

3.3.1 Biological realms

Comparison of terrestrial and marine realms provides perhaps the best example where relative richness is not replicated at different taxonomic levels. Many more animal phyla are known from marine systems than from terrestrial (May 1994). This is also true when we examine lower taxa such as classes (Nicol

1971). Whereas on land we encounter few animal body plans (phyla), with very many variations on a theme (principally insects), all animal phyla with the exception of the Onychophora (or velvet worms) occur in the sea. Furthermore, two-thirds of all animal phyla are exclusively marine (or nearly so). However, fewer than 15% of species currently named are marine, despite the vastly greater area covered by the oceans (May 1994). The extent to which these proportions would change if all of the marine and terrestrial species had been described remains controversial (see Section 2.4). Nevertheless, it seems unlikely that anything like parity with the terrestrial system would be achieved.

May (1994) lists five factors that might help to explain the contrast in the diversities of land and sea:

1 life began in the sea;
2 continental environments are more heterogeneous than marine ones;
3 the ocean-bed environment is less 'architecturally elaborate' than the terrestrial environment;
4 patterns of herbivory differ between land and sea;
5 there are differences in the body size distributions of marine and terrestrial species assemblages.

> ## Going Further
>
> *On your own*
> - Examine the evidence for each of May's (1994) 'five factors'.
> - Decide if, how and why you could say which is most important.
> - How would you justify your decision(s)?

3.3.2 Biogeographic regions

Moving down the spatial scales, there have been numerous attempts to divide the surface of the Earth into broad biogeographic regions, which distinguish areas of marked dissimilarity in the composition of their biota. Comparison of the relative biodiversity of these regions gives a broad-scale picture of its spatial variation. The exercise is somewhat hampered by the variety of existing schemes for the delimitation of these regions. None the less, some important generalities have begun to emerge. First, of the six or so regions commonly recognized in the terrestrial realm, the three 'tropical' regions (Neotropics, Indotropics, Afrotropics) perhaps contain two-thirds or more of all extant terrestrial species. Second, the Neotropics is generally recognized to be the region that contains the greatest overall levels of terrestrial biodiversity. Third, the three 'tropical' regions tend to decline in overall biodiversity from the Neotropics to the Indotropics to the Afrotropics. Levels of biodiversity in the first two are probably more similar to one another, with the Afrotropics being relatively less diverse. In part, this is because the tropical forest of Africa is not as exten-

Table 3.1 Percentages of swallowtail and dragonfly species in different biogeographic regions. (After Sutton & Collins 1991.)

Biogeographic region	Swallowtails	Dragonflies
Indotropical	40	26
Neotropical	29	27
Palaearctic	11	12
Afrotropical	9	16
Oceania	8	11
Nearctic	3	8

sive, well developed or rich as that in the other two regions. Fourth, patterns in the biodiversity of different biogeographic regions may not be consistent amongst many groups of organisms (Table 3.1). For example, the distribution of butterfly species richness amongst regions appears to be more similar to that of birds than of mammals (Robbins & Opler 1997).

In the marine realm, although there are perhaps not so many physical barriers as on land this does not mean that the environment is homogeneous. Indeed, it is made more complex and three-dimensional than the terrestrial realm by the fact that biological life is found at all depths, from the marine intertidal down to about 11 km. This complexity has resulted in there being no agreed classification of regions comparable to that used for the terrestrial realm. Currently, the main biogeographic regions are described with reference to fairly well defined temperature regimes (Couper 1983) or by surface currents (Hayden *et al.* 1984), to give two methods, and not by biological considerations *per se*. Interestingly, of all the species that live in the sea only about 2% live in mid-water, the remainder living on, or in, the sea bed. It has been estimated that up to one-quarter of all marine species and one-fifth of known marine fish species live in coral-reef ecosystems. Reef biodiversity itself is highest in the Indo-Western Pacific, which is also thought to have the world's highest overall marine biodiversity (World Resources Institute 1996). Briggs (1996) used data for echinoderms, molluscs, some crustaceans, reef corals, and fish to show that shelf faunas belonging to the four great tropical regions increased in diversity in the sequence: eastern Atlantic, eastern Pacific, western Atlantic, Indo-Western Pacific.

Up until very recently, the diversity of the deep sea in general was considered to be poor. Even at hydrothermal vent sites, where population density and endemism are found to be high, species richness is still comparatively low (236 recorded species; Tunnicliffe 1991). However, some recent studies on the fauna of deep-sea floors in the Atlantic and Pacific have uncovered a high level of species richness (Grassle 1991; Grassle & Maciolek 1992; Poore & Wilson 1993) which, it could be claimed, may rival even that of coral reefs.

3.3.3 Provinces

The identification of areas of high biodiversity at yet more moderate scales than those of biogeographic regions (here termed provinces) has been a topic of some concern, particularly to conservationists. Most data at these scales tend to refer to countries, whose boundaries often have no biological reality but do reflect an important scale at which many decisions regarding the exploitation and preservation of biodiversity take place (see Chapter 5).

The distribution of biodiversity amongst countries is highly skewed, with a few containing a disproportionately large number of species and most containing a disproportionately small number (Fig. 3.3). Indeed, a set of 'mega-diversity' countries have come to be recognized, comprising the 6–12 countries believed to harbour 50–80% of the world's biodiversity, expressed in terms of species richness (Mittermeier 1988; McNeely *et al.* 1990; Mittermeier & Werner 1990); the list of 12 countries comprises the Malagasy Republic, Australia, China, India, Indonesia, Malaysia, Thailand, Mexico, Brazil, Colombia, Ecuador and Peru. Much of the variation in biodiversity between countries inevitably results from the dramatic differences in their areas, but also reflects such characteristics as their latitude, topographical and habitat diversity, and their human history.

Most countries have rather poor inventories of the flora and fauna that lies within their bounds, let alone details of occurrence. For example, even for those whose faunas have been reasonably well studied, inventories of insect species may remain substantially incomplete (e.g. Japan 29–41% estimated to have been inventoried, Canada 55%, Finland 84%; Gaston 1996a). Moreover,

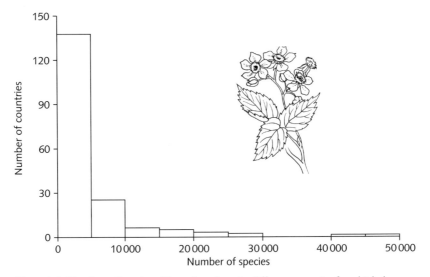

Figure 3.3 Numbers of species of flowering plants in different countries for which data are available (from data in World Conservation Monitoring Centre 1994).

Table 3.2 Estimated numbers of flowering plant species in different regions of the Neotropics. (After Henderson *et al.* 1991.)

Phytogeographic region	Surface area (km²)	Estimated total number of species
Amazon Basin	7 050 000	30 000
Northern Andes	383 000	40 000
Atlantic coastal forest of Brazil	1 000 000	10 000
Central America and Mexico	2 515 295	19 000

the pattern of growth in our knowledge often does not reflect the distribution of biodiversity. Thus, whilst most species occur in the tropics, as many species of insects are presently being described per unit area from temperate regions as from tropical ones (Gaston 1994).

In marine systems, one recently proposed scheme has delineated a number of key (provincial-scale) areas, encompassing mainly coastal locations, out to the rim of the continental shelf (Sherman 1994). These areas (49 in total) are referred to as large marine ecosystems (LMEs). Their importance lies in the fact that 95% of all commercial marine resources are found in these relatively shallow shelf areas (cf. Section 1.4.2, which describes how what you value determines how and what you measure).

The need for explicit quantitative analyses of patterns of biodiversity has been highlighted by the persistence of various misconceptions about its geographical distribution at provincial scales. For example, the overwhelming emphasis on lowland forests arguably has distracted attention from other important patterns in tropical biodiversity. Thus, Henderson *et al.* (1991) show that although the northern Andes has an area only one-twentieth that of the Amazon basin, it contains at least as many species of flowering plants (Table 3.2). Mares (1992) makes a related observation with regard to mammals, for which the Amazon lowlands support fewer species, genera and families than do the drylands of the Neotropics. The drylands are about twice the area of the Amazon lowlands, and on this basis alone such a pattern is not surprising. However, the point remains that the importance of the drylands should not be overlooked, especially as they also contain more endemic species, genera and families. Of course, none of this implies that the Amazon lowlands are not of enormous significance and worthy of the attention they receive.

3.3.4 Endemism

A taxon is *endemic* to an area if it occurs there and nowhere else. The area of endemism can be either relatively large (e.g. the three extant species of monotremes, the echidnas *Tachyglossus aculeatus* and *Zaglossus bruijni* and the

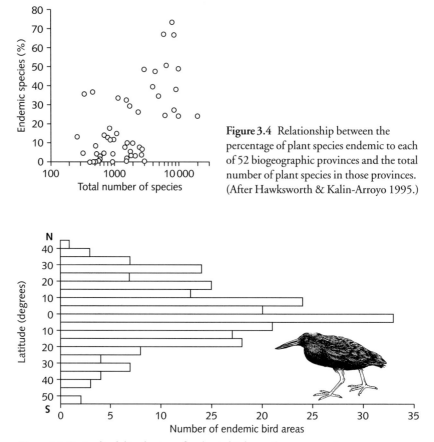

Figure 3.4 Relationship between the percentage of plant species endemic to each of 52 biogeographic provinces and the total number of plant species in those provinces. (After Hawksworth & Kalin-Arroyo 1995.)

Figure 3.5 Latitudinal distribution of endemic bird areas (areas supporting two or more species with geographical ranges of < 50 000 km^2). (After ICBP 1992.)

platypus *Ornithorhynchus anatinus*, are confined to Australia and New Guinea) or very small (e.g. the 'water bear' *Thermozodium esakii* (phylum Tardigrada) is found only in a Japanese hot spring). Some small areas, particularly oceanic islands, can have very high levels of endemism (e.g. Hawaii). On average, however, the proportion of taxa in an area that are endemic to it tends to be an increasing function of the size of the area and the total number of species in that area, though the effects are rather weak (Major 1988; Anderson 1994). For example, the proportion of plant species endemic to each of 52 biogeographic provinces (distributed across all major biomes on all continents) tends to increase with the numbers of plant species that they contain (Fig. 3.4).

More importantly, and more markedly, the number of endemics tends to increase towards lower latitudes. This has been demonstrated in an important study of birds (Fig. 3.5), which has identified 218 endemic bird areas, defined as areas supporting two or more species with restricted ranges (< 50 000 km^2).

In total, these areas occupy a mere 4.5% of the Earth's land surface and contain 73% of all globally threatened bird species; 2649 landbird species (27% of all birds) have breeding ranges of 50 000 km² or less (Long *et al.* 1996).

To some extent, areas of high endemism are also found to be areas of high species richness. For example, Balmford and Long (1995) found that the number of restricted-range bird species in a country increases with the total number of bird species, and Ceballos and Brown (1995) demonstrated that there is a positive correlation across 155 countries between the total number of species of land mammals and the number of these that were endemic.

Going Further

As a group
• Debate the motion: 'If we conserve areas of high species richness (hotspots) we will also conserve all species'. (For organizational details follow the instructions given in the Going Further box on p. 11, Chapter 1.)

References
Prendergast *et al.* (1993), Sisk *et al.* (1994), Williams *et al.* (1996)

3.4 Gradients in biodiversity

3.4.1 Latitudinal gradients in diversity

Perhaps the boldest signature of spatial variation in biodiversity is that associated with latitude. As has long been acknowledged (e.g. Humboldt & Bonpland 1807; Wallace 1853; Bates 1862), the species richness of many groups of organisms increases from high (temperate) to low (tropical) latitudes (Figs 3.6a & 3.7). A similar pattern is also frequently observed for the richness of higher taxa, such as genera (Fig. 3.6b) and families (Fig. 3.8). It is typically manifest whether diversity is determined at local sites (Fig. 3.6), across large regions (Fig. 3.8) or is determined cumulatively across entire latitudinal bands (Fig. 3.7). The steepness of the gradient may vary markedly. Thus, butterflies are more 'tropical' than birds. Although there are approximately two species of butterflies for every species of bird worldwide, birds greatly outnumber butterflies in the Arctic, have about equal numbers of species in temperate North America and are outnumbered by butterflies in the Neotropics (Robbins & Opler 1997).

Terrestrial systems

In the terrestrial realm, the exceptions to the increase in biodiversity towards lower latitudes are relatively scarce. They include a variety of, usually compara-

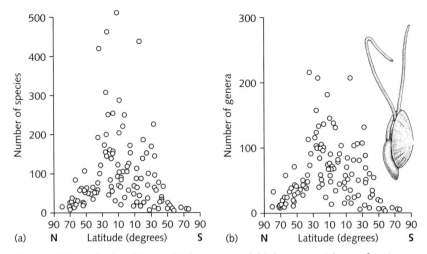

Figure 3.6 Latitudinal gradients in (a) the species and (b) the generic richness of marine bivalves in different localities. (After Flessa & Jablonski 1995.)

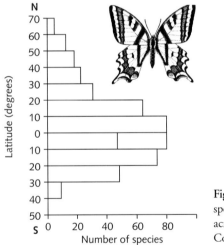

Figure 3.7 Latitudinal gradient in the species richness of swallowtail butterflies across the Americas. (After Sutton & Collins 1991.)

tively minor but sometimes quite major, taxa (Fig. 3.9; e.g. polypore fungi, sawflies, ichneumonid and braconid wasps, aphids), and the patterns of diversity in some regions (most notably Australia). The overall pattern of increasing biodiversity is, however, indisputable.

Marine systems

The question of the generality of latitudinal gradients in diversity in marine systems continues to give rise to some debate. The detection of latitudinal patterns in these environments has been hampered by the confounding effects of

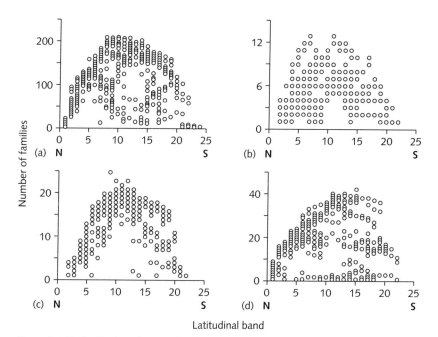

Figure 3.8 Latitudinal gradients in family richness for (a) seed plants, (b) amphibians, (c) reptiles and (d) mammals. Each data point represents the number of families in a cell of a grid of 611 000 km² squares; latitudinal bands run from north (1) to south (24) and the equator lies at the junction of bands 12 and 13. (After Gaston *et al.* 1995.)

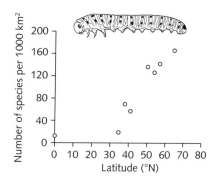

Figure 3.9 Latitudinal gradient in the species richness of sawflies. Each data point is the number of species per 1000 km² in a country or region. (After Kouki *et al.* 1994.)

depth (see Section 3.4.2) and by the problem of attaining any more than a low level of sampling (knowledge of the sediment-dwelling infauna of the deep sea derives from study of less than 2000 quantitative cores, an estimated area of 500 m²; Paterson 1993).

Most contention surrounds patterns in shallow waters. Here, it seems that there are clines of increasing diversity towards lower latitudes for some groups of organisms (Fig. 3.6, Vincent & Clarke 1995), but not for others (e.g. Kendall & Aschan 1993; Lambshead 1993; Dauvin *et al.* 1994; Boucher &

Lambshead 1995; Vincent & Clarke 1995). The pattern, if any, in overall diversity in these systems remains obscure. The numbers of species of perhaps the best-known group, coastal marine fish (teleosts and elasmobranchs), show an increase in richness towards low latitudes (Rohde 1978, 1992). However, amongst coral-reef fish, although clines are reasonably well defined in some biogeographic areas, summing all species occurring in each latitudinal band there is no overall latitudinal gradient in richness (McAllister *et al.* 1994).

In contrast, patterns in the deep sea seem reasonably clear. Rex *et al.* (1993) report latitudinal diversity gradients in the North Atlantic, and strong inter-regional variation in the South Atlantic, for deep-sea bivalves, gastropods and isopods. Poore and Wilson (1993) find a similar pattern for deep-sea isopods. However, some complications seem to remain. Thus, Brey *et al.* (1994) have argued, on the basis of data from the Weddell Sea (Antarctic), that there is no systematic decrease in deep-sea benthic diversity towards polar regions in the Southern Hemisphere.

Pelagic assemblages also appear to exhibit a latitudinal gradient in richness, though again not necessarily a simple one. For example, Angel (1993, 1994a) documents declining species richness towards higher latitudes for ostracods, euphausids, decapods and fish in the water column to a depth of 2000 m at a set of stations in the north-east Atlantic. Such gradients may be stepped rather than smooth, as a result of discontinuities such as the polar front and the sub-tropical convergences (Angel 1994b). McGowan and Walker (1993) argue that the number of species of pelagic plankton is low at high latitudes, but rather than a regular systematic increase toward the equator exhibits a sharp gradient at about 40° N. Diversity is high at mid-latitudes, but in the central and eastern Pacific drops to intermediate levels in the equatorial zone. Diversity increases in the South Pacific and drops to a minimum near Antarctica.

In sum, conflicting evidence and apparently complex patterns in latitudinal clines in the sea mean that these patterns continue to constitute a major challenge to the generality of the statement that diversity increases from temperate to tropical regions.

Large-scale patterns: terrestrial and marine

Where it does occur, two features of the latitudinal gradient of increasing bio-diversity towards lower latitudes are of note. First, it has been a persistent feature of the history of biodiversity. This has been elegantly demonstrated for angiosperms by Crane and Lidgard (1989), who have shown that the pattern was maintained throughout much of the Cretaceous (Fig. 3.10). Second, the gradient is commonly, though far from universally, asymmetrical about the equator. That is, the pattern of diversity across the Earth is more like a pear (increasing rapidly from northern regions to the equator and declining slowly from the equator to southern regions) than an egg (Platnick 1991, 1992). This is well illustrated by the numbers of genera of termites (Fig. 3.11).

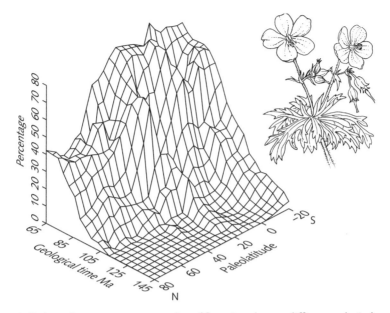

Figure 3.10 Estimated percentage representation of flowering plants at different geological times and at different palaeolatitudes within Cretaceous palynofloras. (After Crane & Lidgard 1989.)

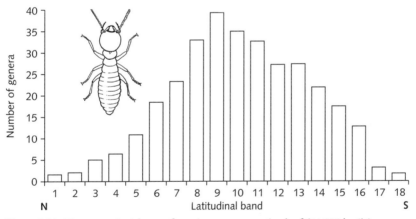

Figure 3.11 Mean generic richness of termites across areas (each of 611 000 km²) in different latitudinal bands (the equator lies at the junction of bands 9 and 10). (After Eggleton 1994.)

Mechanisms

A large number of possible mechanisms for latitudinal gradients in biodiversity has been proposed (Pianka 1966; Stevens 1989; Rohde 1992; Colwell & Hurtt 1994; Rosenzweig 1995; Turner *et al.* 1996). These include the effects of competition, mutualism, predation, patchiness, environmental stability, environmental predictability, productivity, area, number of habitats, ecological time, evolutionary time and solar energy (Rohde 1992). No consensus view on the cause of the pattern seems to be emerging.

Nevertheless, spatial variation in biodiversity is ultimately a product of patterns in rates of origination, immigration, extinction and emigration. At large spatial scales it will tend solely to be a product of origination and extinction. The tropics have thus variously been argued to represent a 'cradle of diversity' exhibiting high origination rates, a 'museum of diversity' with low extinction rates, or some combination of the two. Jablonski (1993), in an analysis of post-Palaeozoic marine orders, has found that there have been significantly more first appearances in tropical waters, whether defined latitudinally or biogeographically, than expected from sampling alone. This provides direct evidence that tropical regions have been a major source of evolutionary novelty.

Diversity and developing countries

The occurrence of most terrestrial biodiversity at low latitudes means that it tends to occur in developing countries (as measured by per capita gross national product, GNP; Fig. 3.12), which obviously has significant implications for the ease with which biodiversity can be maintained (a point we will return to in Chapter 5).

Going Further

On your own or as part of a group
• Gaston (1996b) concludes 'The latitudinal gradient remains a challenge to our understanding of biodiversity, but the perplexity it engenders is perhaps beginning to diminish'. Do you agree with this statement?
• Use this question, and the references given, as the basis for an informal discussion.

References
Stevens (1989), Rohde (1992), Kaufman (1995), Rosenzweig (1995)

3.4.2 Altitudinal and depth gradients in diversity

When we considered species–area relationships (Section 3.2.1) and latitudinal gradients in biodiversity (Section 3.4.1), very little allowance was made for the fact that, quite literally, the Earth is not flat; its surface, above and below water, is moulded into mountains and valleys both by local and global geological processes. Whilst for some purposes it may be useful to refer to the

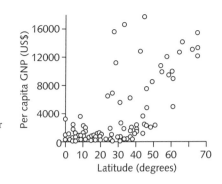

Figure 3.12 Latitudinal gradient in the per capita gross national product (GNP) of countries of the world in 1986. (After Huston 1994.)

Earth's surface using measures of area, the three-dimensional structure of landscapes and seascapes means that sometimes we should really be dealing in volumes.

Altitude

In the terrestrial realm, the third dimension is commonly construed as the altitude of land. Altitude could arguably be ignored when considering large areas, because its magnitude is small compared with those of longitude or latitude. However, it must be remembered that a moderate increase in altitude has, for example, an associated temperature change corresponding to a latitudinal change perhaps of several hundred kilometres.

In terrestrial systems, it is generally accepted that species richness declines with increasing elevation. However, the details of this pattern are variable, with some groups apparently showing a simple decline, whilst others show a hump-shaped relationship in which richness at first increases from low to mid elevations and then declines towards high elevations (although, even here, diversity at low elevations typically exceeds that at high ones). Some of this variation in outcomes may be explained by the differences in area at different elevations, combined with the species–area relationship. For example, Rahbek (1995) has shown that when data are not standardized for differences in area then South American tropical landbirds exhibit a steady decline in richness with elevation, but when these same data are standardized for area a hump-shaped pattern emerges (Fig. 3.13).

Below the Earth's surface

Life occurs beneath the Earth's surface as well as above it, for example in caves at different depths. We know little of the effect of this 'depth gradient' on biodiversity. Certainly the recent exciting discovery of endemic cave communities reliant on chemosynthetic (as opposed to photosynthetic) energy production, similar in function to those occurring in the deep sea at hydrother-

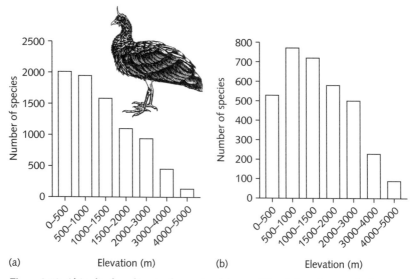

Figure 3.13 Altitudinal gradient in the species richness of South American tropical landbirds, based on data (a) not standardized for elevational variation in area and (b) standardized for such variation. (After Rahbek 1995.)

mal vent sites, is likely to prove of tremendous interest (Sarbu *et al.* 1996). However, life also occurs at even greater depths. Bacterial assemblages have been recovered from up to 4000 m underground, which has been noted as a cause of some concern (we were relieved to read) regarding safety in the development of underground repositories for nuclear waste (Pedersen 1993). While their 'species' richness is not related to depth, such assemblages can consist of up to 62 different 'types' at any one depth (Flierman & Balkwill 1989).

Depth

In some sense, depth can be regarded as the marine equivalent of altitude. However, plainly there are limitations to the parallel because few species are able to achieve a purely aerial existence, and distinction must therefore be drawn between the effects of depth on benthic and pelagic assemblages.

In the pelagic and benthic realms generally, diversity tends to peak at intermediate depths; species richness peaks at depths of 1000–1500 m for pelagic assemblages, and in many taxa increases with increasing depth to a maximum at 1000–2000 m for megabenthos and 2000–3000 m for macrobenthic infauna (Fig. 3.14; Rex 1981; Etter & Grassle 1992; Angel 1993, 1994b) (but compare final part of Section 3.3.2).

The interplays between the various spatial patterns are important in generating the global landscape of biodiversity which we observe. Macpherson and

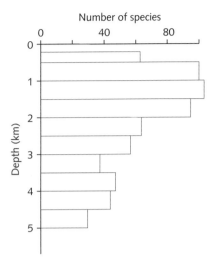

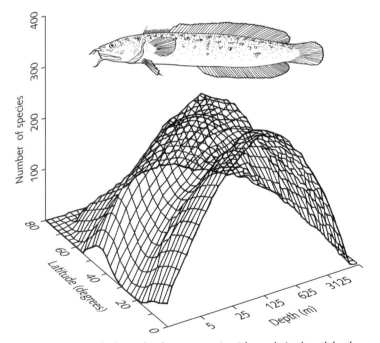

Figure 3.14 Depth gradient in the species richness of megabenthos (summing fish, decapods, holothurians and asteroids) in the Porcupine Seabight region to the south-west of Eire. (After Angel 1994b.)

Figure 3.15 Three-dimensional relationship between species richness, latitude and depth for benthic fishes in the eastern Atlantic. (After Macpherson & Duarte 1994.)

Duarte (1994) examined the effect of both depth and latitude on (amongst other things) the species richness of benthic fish (Fig. 3.15). They found that species richness declined towards higher latitudes (see Section 3.4.1) but at most latitudes species richness also varied with depth; species richness tended to peak at depths of 150–300 m.

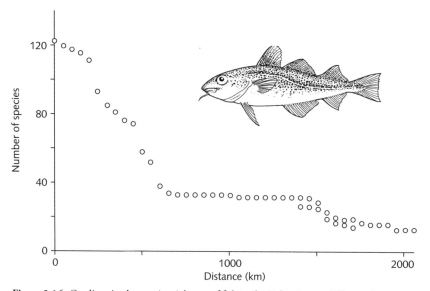

Figure 3.16 Gradient in the species richness of fish in the Baltic Sea, at different distances from the Atlantic–Skagerrak mouth. (After Rapoport 1994.)

3.4.3 Peninsulas and bays

The shapes of landmasses and water bodies can have profound effects on the levels of biodiversity associated with them (by affecting environmental conditions and likelihoods of colonization and extinction), leading to gradients in that diversity. Thus, although not invariable, species richness is often observed to decline towards the tips of peninsulas (the 'peninsula effect') and across bays with distance from the open sea (the 'bay effect') (Fig. 3.16).

3.5 Diversity and environmental variables

Latitude, altitude, depth and land shape do not determine levels of biodiversity *per se*. Rather the gradients in biodiversity associated with these variables result from other changes associated with them, including changes in many environmental variables, although one must be wary of associating observed patterns with causality without further justifications for doing so.

The levels of biodiversity in different areas tend to correlate with the environments of those areas (Fig. 3.17). Analyses have been carried out examining a wide range of parameters (e.g. evapotranspiration, hours of sunshine, precipitation, primary productivity, soil fertility, temperature). The particular relationships observed, however, vary widely between taxonomic groups, regions, and even time of year. Moreover, where they do occur, they need not always be simple; species richness may often peak at some intermediate point along

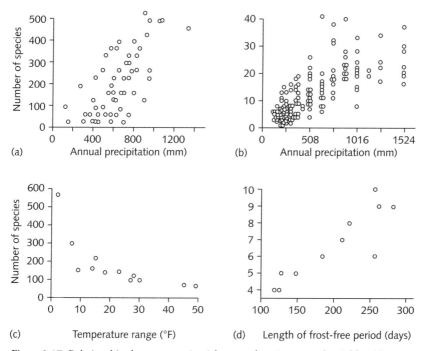

Figure 3.17 Relationships between species richness and environmental variables: (a) woody plants and annual precipitation in areas of southern Africa; (b) frogs and annual precipitation in areas of Australia; (c) birds and range of monthly mean temperatures at sites along the west coast of North America; and (d) lizards and average length of the frost-free period in North American flatland deserts. (a, after O'Brien 1993; b, after Pianka & Schall 1981; c, after MacArthur 1975; d, after Pianka 1967.)

whatever environmental continuum is under investigation.

In general, those environmental factors related to the supply of usable energy (food or limiting nutrient availability, productivity) explain more variation in biodiversity than do those which are not (Wright *et al.* 1993). However, the relationship between energy and species richness tends to be scale dependent. On the global or large regional scale, biodiversity broadly increases with energy, whilst at smaller scales it is a peaked function of energy (Fig. 3.18). The reason for this difference is unclear.

The impact of environmental variables on levels of biodiversity is often demonstrated when human activities change the state of those variables (e.g. by introducing a pollutant). This is dramatically illustrated in Fig. 3.19. In many ways the challenge of the next few years will be the ability to predict and to follow anthropogenic effects, in terms of altered environmental variables (e.g. increases in temperature, chemical pollution), that are superimposed on many of the spatial patterns we document above.

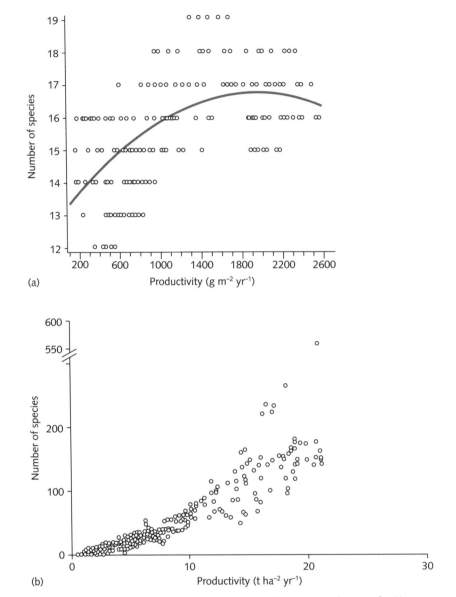

Figure 3.18 Relationships between species richness and primary productivity for (a) mammalian carnivores in areas (63.9 × 63.9 km) of Texas and (b) trees in areas (*c.* 72 000 km²) of America and East Asia. (a, after Owen 1988; b, after Adams & Woodward 1989.)

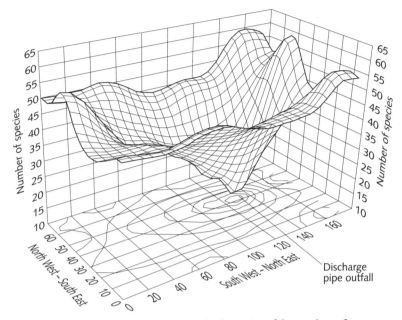

Figure 3.19 A three-dimensional plot showing the depression of the numbers of microscopic sediment-dwelling species of meiofauna associated with an industrial effluent discharged to the outer reaches of a British estuary. The area shown is approximately 1.7 × 1.7 km. (After Anon. 1994.)

3.6 Congruence

If the above patterns constitute the skeleton of an atlas of biodiversity, how do we go about putting the bones together? So we come to perhaps the most fundamental problem in any attempt to develop an atlas of biodiversity, whose resolution is rapidly becoming paramount in biodiversity research. This is, what is the extent of congruence in the patterns of spatial variation in the diversities of different groups of organisms? In other words, to what extent does the diversity of one taxon (say, corals) predict the diversity of other taxa (say, coral-reef fish and crustaceans)? As already highlighted in Section 3.1, given that we know so little about spatial patterns in the biodiversity of highly speciose groups (e.g. bacteria, fungi, insects), if we are to determine patterns in overall biodiversity we need to know to what extent the patterns of diversity of the taxa for which we have such information tell us about the patterns of diversity of these groups. The ease and rapidity with which we can generate an atlas of biodiversity will obviously be greatly enhanced if the patterns are similar.

A starting point in addressing this issue is to ask to what extent the diversities of the different groups about which we do know something themselves tend to be congruent. In a very general sense, of course, we know that many major taxonomic groups show some similarity in their patterns of biodiversity, for

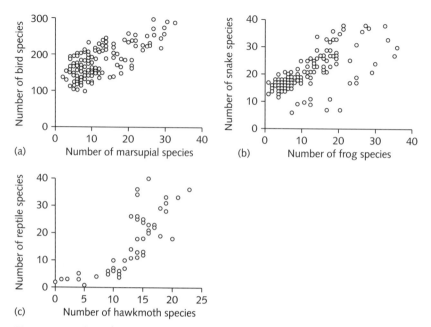

Figure 3.20 Relationships between numbers of (a) bird and marsupial species in 150×150 mile ($\sim 240 \times 240$ km) squares across Australia (from data in Pianka & Schall 1981); (b) snake and frog species in 150×150 mile ($\sim 240 \times 240$ km) squares across Australia (from data in Pianka & Schall 1981); and (c) reptile and hawkmoth species in 152 000 km² squares across Europe.

example in the terrestrial realm they tend to be species-poor in deserts, Arctic regions and at high altitudes, and species-rich in tropical forests and at lower altitudes. However, these are only very rough and ready assertions, and some more refined analyses are required. Unfortunately the results of more detailed studies are mixed and not wholly encouraging. For example, Prendergast *et al.* (1993) examined the coincidence in hotspots of species richness of five groups of animals and plants (birds, butterflies, dragonflies, liverworts, aquatic plants) at the scale of 10×10 km squares across Britain. These groups exhibited only weak coincidence of hotspots and little overlap between coldspots. Whether for hotspots or for covariation in levels of diversity more generally, patterns may be stronger at coarser spatial resolutions (Fig. 3.20). However, even then they may not be consistent (the full range of relationships have been documented, from the strongly positive, through the weak and non-existent, to the strongly negative; Gaston & Williams 1996) and we do not have a good understanding of the frequency of different outcomes.

Where congruences do occur, four sets of mechanisms can potentially explain relationships between the biodiversity of different groups in different areas (Gaston 1996c).

1 Chance: because there is greater overall diversity in some areas, these are

also the most likely areas to contain high biodiversity of individual groups of organisms.

2   Interactions: relationships may result from trophic, competitive or mutualistic interactions between taxa.

3   Common determinants of diversity: relationships may result because the biodiversity of different groups responds in a similar way to variation in a third variable (e.g. temperature, precipitation).

4   Correlated determinants of diversity: relationships may result because the biodiversity of different groups responds to variation in different variables, but these variables are themselves correlated.

Whilst different well-known taxonomic groups may not show strong, or consistent, congruence in biodiversity, the summed biodiversity of these groups may perhaps be taken as the best approximation of the spatial distribution of overall biodiversity that we presently have. Williams *et al.* (1997), taking this approach, have determined the distribution across the Earth of the family richness of terrestrial and freshwater seed plants, amphibians, reptiles and mammals. Although different methods of weighting the families in each group were used (because families may not be comparable between groups; see Section 1.3), the broad pattern was surprisingly robust. Simply summing the numbers of families revealed a hotspot of maximum richness in central Colombia, other hotspots in Nicaragua, Oaxaca (Mexico), southern Colombia and southern peninsular Malaysia, and a strong latitudinal pattern of richness (Fig. 3.21).

3.7 Conclusion

If we cannot be certain how many species there are living on the Earth and we have not described the vast majority of them (see Chapter 2), then mapping

---

Going Further

On your own or as part of a group
• As part of a team of experts investigating the biodiversity of a region you are asked to identify 'candidates' for a good indicator group.
• What are the characteristics you would look for in an indicator group and why?
• Put together a 10-minute oral presentation, assuming that your target audience is political decision-makers.

References
Pearson and Cassola (1992), Pearson (1994), Beccaloni and Gaston (1995), Gaston (1996d)

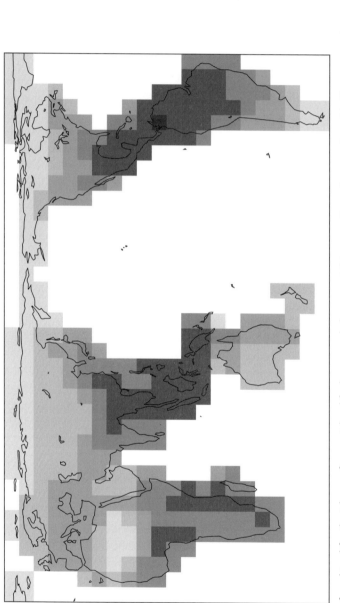

Figure 3.21 Map of combined family richness of terrestrial and freshwater seed plants, amphibians, reptiles and mammals worldwide on an equal area grid of cells of 611 000 km². (After Williams *et al.* 1997.)

the distribution of biodiversity plainly constitutes an enormous challenge. However, some of the major patterns of diversity (latitudinal, altitudinal and depth) are becoming clearer even if the mechanisms are still not fully (or sometimes even partially) understood. Unfortunately, it has only proved possible to extrapolate spatial patterns of well-known groups to biodiversity more broadly in a very general way. Much work is still required to establish spatial patterns and their underlying mechanisms, to identify congruences and, perhaps most critically (and most problematic) of all, to allow us to predict the (continuing) impact of humankind on current patterns of diversity.

3.8 Summary

1   We do not have an atlas that describes exactly how biodiversity is distributed globally, but we are beginning to document many patterns of spatial variation (particularly in species richness).

2   On average, as the size of a geographical area increases so too does the number of species it contains (the species–area relationship).

3   Local species richness tends to be positively correlated with regional species richness.

4   In the terrestrial realm there are few phyla, with many variations on a theme (mainly insects). In the marine realm all but one of the animal phyla occur, but only 15% of all named species.

5   The terrestrial realm can be split into a number of biogeographic regions in a way that is not possible for the marine realm. The tropical regions contain at least two-thirds of all extant terrestrial species, with the Neotropics containing the greatest overall levels of terrestrial biodiversity.

6   At least one-quarter of all marine species live in coral-reef ecosystems, with reef biodiversity and highest marine biodiversity being present in the Indo-western Pacific. The deep sea may be far more diverse than was previously thought.

7   The distribution of terrestrial biodiversity between provinces is uneven, with 6–12 'mega-diversity' countries possessing 50–80% of the world's species.

8   The number of endemics in areas tends to increase towards lower latitudes and high endemism is often found in areas of high species richness.

9   Species richness of many terrestrial and some marine groups increases from temperate to tropical latitudes, although there is no consensus on the mechanisms underlying this pattern. Such latitudinal gradients have been a persistent feature through geological time and are commonly asymmetrical about the equator.

10   In the terrestrial realm, species richness declines with increasing altitude or peaks at intermediate elevations, whilst in the marine realm the relationship with depth is typically hump-shaped.

11 Levels of biodiversity in an area can be correlated with differences (natural and anthropogenic) in environmental factors, although the exact relationships vary widely.

12 Determining the extent of congruence in the pattern of spatial variation in the diversity of different groups of organisms is one possible way in which, with ease and rapidity, we could generate an atlas of biodiversity. Unfortunately, attempts to establish such congruencies have proved disappointing.

3.9 Making connections

The heterogeneous spatial distribution of biodiversity means that some of us encounter more of it, on a day-to-day basis, than do others. But does biodiversity in any sense matter? Having established in this and the previous chapter the essential dynamics of biodiversity, in the next we move on to consider the answer to this important question.

3.10 References

Adams, J.M. & Woodward, F.I. (1989) Patterns in tree species richness as a test of the glacial extinction hypothesis. *Nature* **339**, 699–701.

Anderson, S. (1994) Area and endemism. *Quarterly Review of Biology* **69**, 451–471.

Angel, M.V. (1993) Biodiversity of the pelagic ocean. *Conservation Biology* **7**, 760–772.

Angel, M.V. (1994a) Long-term, large-scale patterns in marine pelagic systems. In: *Aquatic Ecology: Scale, Pattern and Process* (eds P.S. Giller, A.G. Hildrew & D.G. Raffaelli), pp. 403–439. Blackwell Science, Oxford.

Angel, M.V. (1994b) Spatial distribution of marine organisms: patterns and processes. In: *Large-scale Ecology and Conservation Biology* (eds P.J. Edwards, R.M. May & N.R. Webb), pp. 59–109. Blackwell Science, Oxford.

Anon. (1994) *Biodiversity: the UK Action Plan.* HMSO, London.

Balmford, A. & Long, A. (1995) Across-country analyses of biodiversity congruence and current conservation effort in the tropics. *Conservation Biology* **9**, 1539–1547.

Bates, H.W. (1862) Contributions to an insect fauna of the Amazon valley. Lepidoptera: Heliconidae. *Transactions of the Linnean Society* **23**, 495–566.

Beccaloni, G.W. & Gaston, K.J. (1995) Predicting the species richness of Neotropical forest butterflies: Ithomiinae (Lepidoptera: Nymphalidae) as indicators. *Biological Conservation* **71**, 77–86.

Boucher, G. & Lambshead, P.J.D. (1995) Ecological biodiversity of marine nematodes in samples from temperate, tropical, and deep-sea regions. *Conservation Biology* **9**, 1594–1604.

Brey, T., Klages, M., Dahm, C., Gorny, M., Gutt & J., Hain *et al.* (1994) Antarctic benthic diversity. *Nature* **368**, 297.

Briggs, J.C. (1996) Tropical diversity and conservation. *Conservation Biology* **10**, 713–718.

Caley, M.J. & Schluter, D. (1997) The relationship between local and regional diversity. *Ecology* **78**, 70–80.

Ceballos, G. & Brown, J.H. (1995) Global patterns of mammalian diversity, endemism, and endangerment. *Conservation Biology* **9**, 559–568.

Colwell, R.K. & Hurtt, G.C. (1994) Nonbiological gradients in species richness and a spurious Rapoport effect. *American Naturalist* **144**, 570–595.

Cornell, H.V. (1993) Unsaturated patterns in species assemblages: the role of regional processes in setting local species richness. In: *Species Diversity in Ecological Communities: Historical and Geographical Perspectives* (eds R.E. Ricklefs & D. Schluter), pp. 243–252. Chicago University Press, Chicago.

Cornell, H.V. & Lawton, J.H. (1992) Species interactions, local and regional processes, and limits to the richness of ecological communities: a theoretical perspective. *Journal of Animal Ecology* **61**, 1–12.

Couper, A. (ed.) (1983) *The Times World Atlas of the Oceans*. Time Books, London.

Crane, P.R. & Lidgard, S. (1989) Angiosperm diversification and paleolatitudinal gradients in Cretaceous floristic diversity. *Science* **246**, 675–678.

Dauvin, J.-C., Kendall, M., Paterson, G., Gentil, F., Jirkov, I., Sheader, M. & De Lange, M. (1994) An initial assessment of polychaete diversity in the northeastern Atlantic Ocean. *Biodiversity Letters* **2**, 171–181.

Eggleton, P. (1994) Termites live in a pear-shaped world: a response to Platnick. *Journal of Natural History* **28**, 1209–1212.

Etter, R.J. & Grassle, J.F. (1992) Patterns of species diversity in the deep sea as a function of sediment particle size diversity. *Nature* **360**, 576–578.

Flessa, K.W. & Jablonski, D. (1995) Biogeography of recent marine bivalve molluscs and its implications for paleobiogeography and the geography of extinction: a progress report. *Historical Biology* **10**, 25–47.

Flierman, C.B. & Balkwill, D.L. (1989) Microbial life in deep terrestrial subsurfaces. *Bioscience* **39**, 370–377.

Gaston, K.J. (1994) Spatial patterns of species description: how is our knowledge of the global insect fauna growing? *Biological Conservation* **67**, 37–40.

Gaston, K.J. (1996a) Species richness: measure and measurement. In: *Biodiversity: a Biology of Numbers and Difference* (ed. K.J. Gaston), pp. 77–113. Blackwell Science, Oxford.

Gaston, K.J. (1996b) Biodiversity – latitudinal gradients. *Progress in Physical Geography* **20**, 466–476.

Gaston, K.J. (1996c) Spatial covariance in the species richness of higher taxa. In: *Aspects of the Genesis and Maintenance of Biological Diversity* (eds M.E. Hochberg, J. Clobert & R. Barbault), pp. 221–242. Oxford University Press, Oxford.

Gaston, K.J. (1996d) Biodiversity – congruence. *Progress in Physical Geography* **20**, 105–112.

Gaston, K.J. & Williams, P.H. (1996) Spatial patterns in taxonomic diversity. In: *Biodiversity: a Biology of Numbers and Difference* (ed. K.J. Gaston), pp. 202–229. Blackwell Science, Oxford.

Gaston, K.J., Williams, P.H., Eggleton, P. & Humphries, C.J. (1995) Large scale patterns of biodiversity: spatial variation in family richness. *Proceedings of the Royal Society of London Series B* **260**, 149–154.

Grassle, J.F. (1991) Deep-sea benthic biodiversity. *Bioscience* **51**, 464–469.

Grassle, J.F. & Maciolek, N.J. (1992) Deep-sea species richness: regional and local diversity estimates from quantitative bottom samples. *American Naturalist* **139**, 313–341.

Griffiths, D. (1997) Local and regional species richness in North American lacustrine fish. *Journal of Animal Ecology* **66**, 49–56.

Hawksworth, D.L. & Kalin-Arroyo, M.T. (1995) Magnitude and distribution of biodiversity. In: *Global Biodiversity Assessment* (ed. V.H. Heywood), pp. 107–199. Cambridge University Press, Cambridge.

Hayden, B.P., Ray, C.G. & Dolan, R. (1984) Classification of coastal and marine environments. *Environmental Conservation* 11, 199–207.

Henderson, A., Churchill, S. & Luteyn, J. (1991) Neotropical plant diversity. *Nature* 351, 21–22.

Humboldt, A. & Bonpland, A. (1807) *Essai sur la Géographie des Plantes Accompagné d'un Tableau Physique des Régions Équinoxiales.* (Reprinted 1977, Arno Press, New York.)

Huston, M.A. (1994) *Biological Diversity: the Coexistence of Species on Changing Landscapes.* Cambridge University Press, Cambridge.

ICBP (1992) *Putting Biodiversity on the Map: Priority Areas for Global Conservation.* ICBP (BirdLife International), Cambridge.

Jablonski, D. (1993) The tropics as a source of evolutionary novelty through geological time. *Nature* 364, 142–144.

Judas, M. (1988) The species–area relationship of European Lumbricidae (Annelida, Oligochaeta). *Oecologia* 76, 579–587.

Kaufman, D.M. (1995) Diversity of New World mammals: universality of the latitudinal gradients of species and bauplans. *Journal of Mammalogy* 76, 322–334.

Kendall, M.A. & Aschan, M. (1993) Latitudinal gradients in the structure of macrobenthic communities: a comparison of Arctic, temperate and tropical sites. *Journal of Experimental Marine Biology and Ecology* 172, 157–169.

Kouki, J., Niemelä, P. & Viitasaari, M. (1994) Reversed latitudinal gradient in species richness of sawflies (Hymenoptera, Symphyta). *Annales Zoologici Fennici* 31, 83–88.

Lambshead, P.J.D. (1993) Recent developments in marine benthic biodiversity research. *Océanis* 19, 5–24.

Long, A.J., Crosby, M.J., Stattersfield, A.J. & Wege, D.C. (1996) Towards a global map of biodiversity: patterns in the distribution of restricted-range birds. *Global Ecology and Biogeography Letters* 5, 281–304.

McAllister, D.E., Schueler, F.W., Roberts, C.M. & Hawkins, J.P. (1994) Mapping and GIS analysis of the global distribution of coral reef fishes on an equal-area grid. In: *Mapping the Diversity of Nature* (ed. R.I. Miller), pp. 155–175. Chapman & Hall, London.

MacArthur, J.W. (1975) Environmental fluctuations and species diversity. In: *Ecology and Evolution of Communities* (eds M.L. Cody & J.M. Diamond), pp. 74–80. Belknap Press, Cambridge, Massachusetts.

MacArthur, R.H. & Wilson, E.O. (1967) *The Theory of Island Biogeography.* Princeton University Press, Princeton.

McGowan, J.A. & Walker, P.W. (1993) Pelagic diversity patterns. In: *Species Diversity in Ecological Communities: Historical and Geographical Perspectives* (eds R.E. Ricklefs & D. Schluter), pp. 203–214. Chicago University Press, Chicago.

McNeely, J.A., Miller, K.R., Reid, W.V., Mittermeier, R.A. & Werner, T.B. (1990) *Conserving the World's Biodiversity.* IUCN, WRI, CI, WWF and The World Bank, Washington, DC.

Macpherson, E. & Duarte, C.M. (1994) Patterns in species richness, size, and latitudinal range of East Atlantic fishes. *Ecography* 17, 242–248.

Major, J. (1988) Endemism: a botanical perspective. In: *Analytical Biogeography: an Integrated Approach to the Study of Animal and Plant Distributions* (eds A.A. Myers & P.S. Giller), pp. 117–146. Chapman & Hall, London.

Mares, M.A. (1992) Neotropical mammals and the myth of Amazonian biodiversity. *Science* 255, 976–979.

May, R.M. (1994) Biological diversity: differences between land and sea. *Philosophical Transactions of the Royal Society of London Series B* 343, 105–111.

Mittermeier, R.A. (1988) Primate diversity and the tropical forest: case studies from Brazil and Madagascar and the importance of the megadiversity countries. In: *Biodiversity* (eds E.O. Wilson & F.M. Peter), pp. 145–154. National Academy Press, Washington, DC.

Mittermeier, R.A. & Werner, T.B. (1990) Wealth of plants and animals unites 'megadiversity' countries. *Tropicus* 4, 4–5.

Nicol, D. (1971) Species, class, and phylum diversity of animals. *Quarterly Journal of Florida Academy of Science* 34, 191–194.

O'Brien, E.M. (1993) Climatic gradients in woody plant species richness: towards an explanation based on an analysis of southern Africa's woody flora. *Journal of Biogeography* 20, 181–198.

Owen, J.G. (1988) On productivity as a predictor of rodent and carnivore diversity. *Ecology* 69, 1161–1165.

Paterson, G.L.J. (1993) *Patterns of polychaete assemblage structure from bathymetric transects in the Rockall Trough, NE Atlantic Ocean.* PhD thesis, University of Wales.

Pearson, D.L. (1994) Selecting indicator taxa for the quantitative assessment of biodiversity. *Philosophical Transactions of the Royal Society of London Series B* 345, 75–79.

Pearson, D.L. & Cassola, F. (1992) World-wide species richness patterns of tiger beetles (Coleoptera: Cicindelidae): indicator taxon for biodiversity and conservation studies. *Conservation Biology* 6, 376–391.

Pedersen, K. (1993) The deep subterranean biosphere. *Earth-Science Reviews* 34, 243–260.

Pianka, E.R. (1966) Latitudinal gradients in species diversity: a review of concepts. *American Naturalist* 100, 33–46.

Pianka, E.R. (1967) On lizard species diversity: North American flatland deserts. *Ecology* 48, 333–351.

Pianka, E.R. & Schall, J.J. (1981) Species densities of Australian vertebrates. In: *Ecological Biogeography of Australia* (ed. A. Keast), pp. 1675–1694. Junk, The Hague.

Platnick, N.I. (1991) Patterns of biodiversity: tropical vs. temperate. *Journal of Natural History* 25, 1083–1088.

Platnick, N.I. (1992) Patterns of biodiversity. In: *Systematics, Ecology, and the Biodiversity Crisis* (ed. N. Eldredge), pp. 15–24. Columbia University Press, New York.

Poore, G.C.B. & Wilson, G.D.F. (1993) Marine species richness. *Nature* 361, 597–598.

Prendergast, J.R., Quinn, R.M., Lawton, J.H., Eversham, B.C. & Gibbons, D.W. (1993) Rare species, the coincidence of diversity hotspots and conservation strategies. *Nature* 365, 335–337.

Rahbek, C. (1995) The elevational gradient of species richness: a uniform pattern? *Ecography* 18, 200–205.

Rapoport, E.H. (1994) Remarks on marine and continental biogeography: an areographical viewpoint. *Philosophical Transactions of the Royal Society of London Series B* 343, 71–78.

Rex, M.A. (1981) Community structure in the deep-sea benthos. *Annual Review of Ecology and Systematics* 12, 331–354.

Rex, M.A., Stuart, C.T., Hessler, R.R., Allen, J.A., Sanders, H.L. & Wilson, G.D.F. (1993) Global-scale latitudinal patterns of species diversity in the deep-sea benthos. *Nature* 365, 636–639.

Ricklefs, R.E. (1987) Community diversity: relative roles of local and regional processes. *Science* 235, 167–171.

Robbins, R.K. & Opler, P.A. (1997) Butterfly diversity and a preliminary comparison with bird and mammal diversity. In: *Biodiversity II* (eds M.L. Reaka-Kudla, D.E. Wilson & E.O. Wilson), pp. 69–82. Joseph Henry Press, Washington, DC.

Rohde, K. (1978) Latitudinal gradients in species diversity and their causes. II. Marine parasitological evidence for a time hypothesis. *Biologisches Zentralblatt* 97, 405–418.

Rohde, K. (1992) Latitudinal gradients in species diversity: the search for the primary cause. *Oikos* 65, 514–527.

Rosenzweig, M.L. (1995) *Species Diversity in Space and Time*. Cambridge University Press, Cambridge.

Sarbu, S.M., Kane, T.C. & Kinkle, B.K. (1996) A chemoautotrophically based cave ecosystem. *Science* 272, 1953–1955.

Saunders, D.A., Hobbs, R.J. & Margules, C.R. (1991) Biological consequences of ecosystem fragmentation: a review. *Conservation Biology* 5, 18–32.

Sclater, W.L. & Sclater, P.L. (1899) *The Geography of Mammals*. Kegan Paul, Trench, Trübner & Co., London.

Sherman, K. (1994) Sustainability, biomass yields, and health of coastal ecosystems: an ecological perspective. *Marine Ecology: Progress Series* 112, 277–301.

Sisk, T.D., Launer, A.E., Switky, K.R. & Ehrlich, P.R. (1994) Identifying extinction threats. *BioScience* 44, 592–604.

Stevens, G.C. (1989) The latitudinal gradient in geographical range: how so many species coexist in the tropics. *American Naturalist* 133, 240–256.

Sutton, S.L. & Collins, N.M. (1991) Insects and tropical forest conservation. In: *The Conservation of Insects and their Habitats* (eds N.M. Collins & J.A. Thomas), pp. 405–424. Academic Press, London.

Tunnicliffe, V. (1991) The biology of hydrothermal vents: ecology and evolution. *Oceanography and Marine Biology Annual Review* 29, 319–407.

Turner, J.R.G., Lennon, J.J. & Greenwood, J.J.D. (1996) Does climate cause the global biodiversity gradient? In: *Aspects of the Genesis and Maintenance of Biological Diversity* (eds M.E. Hochberg, J. Clobert & R. Barbault), pp. 199–220. Oxford University Press, Oxford.

Vincent, A. & Clarke, A. (1995) Diversity in the marine environment. *Trends in Ecology and Evolution* 10, 55–56.

Wallace, A.R. (1853) On the habits of the butterflies of the Amazon valley. *Transactions of the Entomological Society of London (N.S.)* 2, 253–264.

Williams, P., Gibbons, D., Margules, C., Rebelo, A., Humphries, C. & Pressey, R. (1996) A comparison of richness hotspots, rarity hotspots, and complementary areas for conserving diversity of British birds. *Conservation Biology* 10, 155–174.

Williams, P.H., Gaston, K.J. & Humphries, C.J. (1997) Mapping biodiversity value worldwide: combining higher-taxon richness from different groups. *Proceedings of the Royal Society of London Series B* 264, 141–148.

Williamson, M. (1988) Relationship of species number to area, distance and other variables. In: *Analytical Biogeography: an Integrated Approach to the Study of Animal and Plant Distributions* (eds A.A. Myers & P.S. Giller), pp. 91–115. Chapman & Hall, London.

World Conservation Monitoring Centre (Comp.) (1994) *Biodiversity Data Sourcebook*. World Conservation Press, Cambridge.

World Resources Institute (1996) *World Resources 1996–97*. Oxford University Press, Oxford.

Wright, D.H., Currie, D.J. & Maurer, B.A. (1993) Energy supply and patterns of species richness on local and regional scales. In: *Species Diversity in Ecological Communities: Historical and Geographical Perspectives* (eds R.E. Ricklefs & D. Schluter), pp. 66–74. Chicago University Press, Chicago.

3.11 Further reading

Beddington, J.R., Cushing, D.H., May, R.M. & Steele, J.H. (eds) (1994) Generalising across marine and terrestrial ecology. *Philosophical Transactions of the Royal Society of London Series B* **343**, 1–111. (*The papers from an important discussion meeting.*)

Begon, M., Harper, J.L. & Townsend, C.R. (1996) *Ecology: Individuals, Populations and Communities*. Blackwell Science, Oxford. (*A superb treatment of ecology, including the ecological issues touched on in this chapter.*)

Brown, J.H. (1995) *Macroecology*. University of Chicago Press, Chicago. (*An introduction to macroecology, by its chief proponent.*)

Brown, J.H. & Gibson, A.C. (1983) *Biogeography*. C.V. Mosby, St Louis. (*To our mind unsurpassed as an introduction to biogeography, but now difficult to obtain.*)

Gaston, K.J. (1994) *Rarity*. Chapman & Hall, London. (*A synthesis of what is known about rarity, much of it related to the patterns of biodiversity.*)

Kruckeberg, A.R. & Rabinowitz, D. (1985) Biological aspects of endemism in higher plants. *Annual Review of Ecology and Systematics* **16**, 447–479. (*A useful introduction to endemism, with many good examples.*)

Meadows, P.S. & Campbell, J.I. (1988) *An Introduction to Marine Science*, 2nd edn. Blackie, Glasgow. (*An excellent introduction to a study of the sea. Written in an extremely concise, almost note-like form, it contains much useful background to studying biodiversity in the marine environment.*)

Myers, A.A. & Giller, P.S. (eds) (1988) *Analytical Biogeography: an Integrated Approach to the Study of Animal and Plant Distributions*. Chapman & Hall, London. (*Remains perhaps the best single-volume treatment of many of the primary issues in biogeography.*)

Ricklefs, R.E. & Schluter, D. (eds) (1993) *Species Diversity in Ecological Communities: Historical and Geographical Perspectives*. Chicago University Press, Chicago. (*A landmark text exploring the roles of large-scale spatial and temporal processes in generating and maintaining diversity.*)

Williamson, M. (1981) *Island Populations*. Oxford University Press, Oxford. (*Remains the authoritative statement on biodiversity on islands.*)

4 Does biodiversity matter?

4.1 Introduction

The variety of life is manifestly complex (Chapter 1), has changed dramatically through time (Chapter 2) and is unevenly distributed through space (Chapter 3). For many of us, these observations may be interesting in their own right, and the study of biodiversity may be largely a heuristic exercise. But this ignores a fundamental question that, particularly against a background of unprecedented losses in biodiversity (see Section 2.5), demands both an intellectual and a practical response: Does biodiversity matter?

In attempting to answer this question we discuss the sorts of things that might be, or are, valued about biodiversity and why (using 'value' in the broadest sense, not simply as a shorthand for monetary worth). In the space available, we pass swiftly over many well-trodden, and some less frequented, paths. The order in which these values are presented is not indicative of their relative importance or even an endorsement of their validity (what to one person is 'the legitimate exploitation of natural resources' can be to another 'the rape of the natural world'). Having established that no matter who you are, all of us value biodiversity at some level, and given that we are currently losing biodiversity, we then move on to consider a frequently posed question: How much biodiversity do we need? We conclude by asking if indeed these are the sorts of questions we should be asking, and whether there are not some more pertinent alternatives.

4.2 Use value

The values of biodiversity can be divided into two broad and largely self-explanatory categories, use value and non-use value. These categories are not always clear-cut, particularly when we are discussing the intrinsic value of biodiversity, but they are still helpful as long as we are mindful of their limitations. Use value has two components, direct use value and indirect use value.

4.2.1 Direct use value

Direct use value derives from the direct role of biological resources in consumption or production. It essentially concerns *marketable commodities*. The scale of the exploitation of biodiversity is enormous and multifaceted (for fuller discussion and references see World Conservation Monitoring Centre 1992; Heywood 1995; Kunin & Lawton 1996). However, it remains to be fully evaluated, a task that is perhaps practically impossible.

Types of exploitation

Selected types of the direct use value of biodiversity are discussed under some broad headings.

Food

Biodiversity provides food for humans in forms that include meat, fruit, nuts and vegetables, and adjuncts to food in the form of food colourants, flavourings and preservatives. These may derive from wild or cultivated sources. Of the 250 000 or so species of flowering plants, about 3000 have been regarded as a food source and around 200 have been domesticated for food. However, at present more than 80% of the food supply of the human population is obtained, directly or indirectly, from just 20 kinds of plants. Production of major food crops in 1991 totalled 1890 Mt (World Resources Institute 1994).

The diversity of animals exploited for food is more difficult to enumerate, although again the wide range of species consumed contrasts with most consumption being concentrated on a small proportion of these species. Even so, the vast scale of the exploitation is readily apparent. For example, world landings of aquatic resources totalled 99.5 Mt in 1989, of which almost 70% were used for human consumption (World Conservation Monitoring Centre 1992).

Whether of animals or plants, the diversity of organisms exploited for food is rather narrow when contrasted with their overall diversity, leaving a vast potential for further exploitation. This gap is chiefly being closed indirectly, through the use of wild species and varieties to supply genes for the improvement of cultivated species (increasing yields, tolerances and disease resistance). Indeed, broadening the genetic base of some food species may perhaps be the only way in which our heavy reliance upon them can be maintained.

Medicine

A significant proportion of drugs are derived, directly or indirectly, from biological sources. In the USA, almost one-quarter of all medical prescriptions are for formulations based on plant or microbial products or on derivatives or synthetic versions of them (Eisner 1989). Around 119 pure chemical substances extracted from some 90 species of higher plants are used in medicines throughout the world, and over 21 000 names (including synonyms) are associated with plants which have reported medical uses (WCMC 1992). In all, however, only about 5000, predominantly temperate, species of higher plants, have been thoroughly investigated as potential sources of new drugs (WCMC 1992). Exploration of the medical (as well as other) potential of plants and microorganisms, both those known and those yet to be discovered, is very much a 'growth area'.

Animals also find extensive use in traditional remedies (with international trade in association with Oriental medicine being substantial), as a source of a range of useful products (e.g., anticoagulants, coagulants, vasodilatory agents), and for models on which to test potentially useful drugs or techniques.

Biological control

The use of natural enemies to control species we regard as pests is increasingly widespread. Biocontrol programmes have been attempted against several hundred species of plants and insects, with approximately 30% of weed control and 40% of insect biocontrol programmes being successful (Kunin & Lawton 1996). The economic returns can be huge, with the monetary values of annual gains in food or other crop production perhaps exceeding many times over the entire investment in control programmes. For example, the cost-benefit ratio for the control of cassava mealybug *Phenacoccus manihoti* by the encyrtid wasp *Epidinocarsis lopezi* in Africa was estimated to be 1–149 with annual savings as high as US$250 million (Norgaard 1988).

Industrial materials

A wide range of industrial materials, or templates for the production of such materials, are derived directly from biological resources. These include building materials, fibres, dyes, resins, gums, adhesives, rubber, oils and waxes, agricultural chemicals (including pesticides), and perfumes. For wood alone, in 1989 the total worldwide value of exports was estimated to be US$6 billion (WCMC 1992), and more than 3.8 million cubic metres are estimated to be harvested annually worldwide, for fuel, timber and pulp (Kunin & Lawton 1966). Including agriculture, food processing, industrial chemical and pollution control sectors, the biotechnology industry made sales of US$10–12 billion in 1993 in the United States alone (these are projected to reach US$100 billion by 2035; Colwell 1997).

As is the case for food and medicine, the scope for exploitation of a far greater diversity of organisms for industrial materials is vast. The reasons that it is so much greater than presently realised, probably has as much to do with cultural factors ('the devil you know') as it does with our ignorance of natural products.

Recreational harvesting

Examples of recreational harvesting are multifarious but include hunting and fishing, the harvesting of animals (e.g., fish, reptiles, birds, mammals) for display and as pets, and the harvesting of plants for personal and public gardens (in the British Isles alone, some 25 000 plant species are grown in botanic gardens and 14 000 are commercially available garden plants; Crawley 1997). The annual trade in live wild birds is estimated to have been between 2 and 5 million animals in the 1980s (Thomsen *et al.* 1992). Likewise, there is significant international trade in ornamental plants derived from the wild, including orchids, cacti and other succulents, and bulbs.

Ecotourism

Ecotourism is by definition founded on biodiversity, and has developed into a massive industry. Indeed, tourism as a whole is one of the fastest growing industries in the world. In 1988 an estimated 157–236 million people took part in international ecotourism (i.e., in countries of which they were not nationals), contributing between US$93 and US$233 billion to national incomes (Filion *et al.* 1994). However, international tourism is also estimated to account for perhaps only 9% of global tourism receipts (the rest is domestic), suggesting that these figures present only a fraction of the scale and economic impact of ecotourism (Filion *et al.* 1994).

Desirability and sustainability

Particular forms of exploitation of biodiversity may be regarded by some as distasteful (e.g. recreational hunting). Many acts of exploitation are certainly not sustainable. Some species exploited for food continue to be driven towards extinction, rates of deforestation continue to vastly outstrip forest regeneration, and ecotourism may in some instances result in the emigration or even death of the very organisms which people came to see.

Illustrative of non-sustainable use is the total global fish harvest, which in 1993 was 101 Mt, with more than two-thirds of the world's marine fish stocks estimated as being fished at, or beyond, their level of maximum productivity (World Resources Institute 1994). Likewise, Lean and Hinrichsen (1990) state that almost 1.3 billion people in developing countries are consuming fuelwood faster than it is being replenished and that if present trends continue another one billion will be faced with chronic fuelwood shortages by the turn of the century.

Unfortunately, from a strictly economic standpoint ('knowing the cost of everything and the value of nothing?') non-sustainable use could still be regarded as a 'viable' option in some cases. For example, the 'best' long-term harvesting strategy for biological populations with relatively low growth rates may be to exploit them to extinction. The revenue generated by this harvest when invested could conceivably yield a greater cash return than that generated by the sustainable harvest from the population (Clark 1981; Lande *et al.* 1994; May 1994). However, such a view could only be held by ignoring both the indirect use value and the non-use value of biological resources.

4.2.2 Indirect use value

The indirect use value of biodiversity derives from the many functions it performs in providing services that are crucial to human well-being (Table 4.1; Westman 1977; Ehrlich & Ehrlich 1992). These services can in some sense be regarded as 'free', in that they tend not to be the subject of direct trading in the marketplace. Alongside those perhaps more readily recognized, such as nutrient cycling and soil formation, there are numerous others. For example, many non-commercial species of marine molluscs and crustaceans may not be used directly themselves, but none the less constitute an essential food source for many economically important fish species. The value of these invertebrates is indirect as they derive their value (in an economic sense) from the fish. Likewise, declines in the diversity and numbers of wild bees in many areas (often as a product of habitat destruction) has drawn attention to their agricultural significance as pollinators and to the adverse effects on crop yields of these losses (O'Toole 1993).

Some natural environments have both a direct and an indirect value. Take, for example, a tropical forest. This may provide a number of direct use values, including those of timber, medicinal plants, other forest products, hunting and fishing, recreation and tourism. It may also provide indirect use values, including soil conservation and soil productivity, and watershed protection (with consequences for water supply and storage, flood control, climate and carbon sequestration; Perrings 1995).

Indirect use values are more difficult to quantify or cost than direct use values and in some cases we may not be able to recognize let alone explain them. Nowhere is the need to maintain biodiversity because of the services it provides (even if we don't always understand these) more graphically illustrated than in the recent Biosphere 2 experiments (Cohen & Tilman 1996). Biosphere 2 is the world's largest closed-environment facility, a 1.27 ha (3.15 acre) area, containing soil, air, water, plants and animals. Roughly US$200 million were invested in its design and construction, millions more in its operation (annual energy investments exceeded US$1 million) and it could draw (and has drawn) on immense technological resources and expertise. None the less, it

Atmospheric 'regulation'
Climatic 'regulation'
Hydrological 'regulation'
Nutrient cycling
Pest control
Photosynthesis
Pollination
Soil formation and maintenance

Table 4.1 Some ecosystem services provided by biodiversity.

proved impossible to create a materially closed system that could support eight humans with adequate food, water and air for two years. Surprise changes in the environment included a dramatic fall in oxygen levels and rise in carbon dioxide, a rise in N_2O concentrations, overloading of water systems with nutrients, and the extinction of all pollinators. In short we, with all our technology, ingenuity and 'unlimited' (compared with normal science budgets) financial resources, cannot build systems that will provide eight humans, let alone humankind, with the life-supporting services that natural ecosystems provide us for free.

Going Further

On your own or as part of a group
• How much biodiversity do you use in a week?
• If you are going to attempt this as a group activity, you should first appoint a chairperson. The chairperson is then responsible for (a) asking the members of the group to make their own individual list, (b) leading a group discussion based on the contents of these lists, and (c) drawing the discussion to an end by bringing out common themes touched upon by most of the group members.

4.3 Non-use value

> *I am delighted to know that no final decision on the colour [of a new London bus] has been made and hope it will not be determined by finance alone. To be guided in these matters by money values means really to be guided by no values at all.* Sir Hugh Casson, letter to *The Times* (20 September 1963)

Non-use value is that associated with biological resources even if they are not directly or indirectly exploited. Non-use value can be divided into at least three components: option value, bequest value and intrinsic value.

4.3.1 Option and bequest values

In addition to wishing biodiversity to be maintained for its current direct and indirect use value, we may be interested in retaining biodiversity for the options for future use or non-use that it provides (Weisbrod 1964). For example, the importance cannot be overstated of maintaining genetic diversity both in terms of the evolutionary potential of a species as its environment changes and as a source of 'raw material' for the selection of desirable genetic features in the future. These option values may also include the knowledge (of practical or heuristic significance) embodied in organisms, in as much as the loss of a species represents the loss of information (Morowitz 1991).

Once extinct, a species cannot be recovered and particular species currently of no known value could well prove invaluable in the future. The relative importance even of known species, or particular use values, may shift and change over a time-scale of generations or less. By preserving species or setting up conservation strategies now, we give future generations the option to retain and value what we perhaps did not value enough. This takes us on to bequest value or the value of passing on a resource, in this case biodiversity, intact (or as near as possible) to future generations (Krutilla 1967).

The philosopher John Locke suggested that each generation should bequeath 'enough and as good for others' to future generations not just because they should, but because justice demands it. The modern version of this is the slightly more elaborated 'justice as opportunity' view that says we should also 'compensate' our children in the future for the loss of wealth, production or ecosystem services we are currently responsible for ('leaving the world in a better state than you found it in'). This notion is embodied in the final section of the Preamble to the Convention, which states that contracting parties are:

> . . . *determined to conserve and sustainably use biological diversity for the benefit of present and future generations.*

4.3.2 Intrinsic value

Up to this point all of the arguments we have considered for the importance of biodiversity have been based, in one way or another, on marketable commodities and non-market goods and services. They assume that 'value', or whether or not biodiversity 'matters', is solely expressed in terms of the well-being of humanity. However, organisms may equally be seen as having intrinsic value, irrespective of the uses to which they may or may not be put. The debate is made complicated by the fact that it is often difficult to disentangle and/or recognize the different values present. For example, apart from those referred to above, the services to humankind provided by biodiversity may also include the pleasure derived from interacting with wild organisms directly, or indirectly via the cultural activities they inspire (paintings, literature, theatre, etc.) and from the knowledge that these organisms exist (their existence value, even if they are never actually seen by the individual). Even if it is difficult to categorize these different facets, few would deny that there is much of value here. In fact much of human culture, history and belief has been, and is, inextricably tied up with biodiversity, often in ways that we cannot begin to comprehend.

There is a heavily reductionist, or materialistic, world view that attempts to explain concern for the environment as thinly veiled self-interest or merely as a response to peer pressure. While such a view is not without its influential proponents, many others from both secular and religious standpoints maintain that people have an absolute moral responsibility to protect what are our only known living companions in the universe (Ehrlich & Wilson 1991). They would reject the notion that predicates of value are merely statements of self-interest or subjective feeling.

Wilson (1993) believes that humankind, as 'bearers of life' themselves, recognize and have empathy with other 'bearers of life' and this naturally predisposes them to an appropriate care of, and for, life in all its multifaceted forms. The opening section of the Preamble to the Convention recognizes the:

> *intrinsic value of biological diversity and of the ecological, genetic, social, economic, scientific, educational, cultural, recreational and aesthetic values of biological diversity and its components.*

Perhaps it is telling that intrinsic value is placed before, and apart from, the other values. The notion of biodiversity having intrinsic value is also found in other treaties, both regional (e.g. Convention on the Conservation of European Wildlife and Natural Habitats 1979, Berne) and global (e.g. World Charter for Nature 1982).

To varying degrees the ethic of the intrinsic worth of the natural world is rooted in most religions. This is particularly so in many of the Asian religions such as Buddhism, Hinduism and Jainism where humans are considered to be an integral part of the natural world. It has been claimed that the major religions that set humankind apart from, and 'above', the natural world, namely Christianity, Islam and Judaism, have in so doing promoted a world view that places little intrinsic value on biodiversity except as a resource to be exploited (White 1967). Historically there is much truth in this claim (noting, however, that exploitation of biodiversity was not absent from countries with alternative and more 'biodiversity-friendly' world views), although the basis for a respect for biodiversity itself can be found in the scriptures of these religions. For example, while the Judaic Scriptures (the Christian Old Testament) paint a picture of humankind as distinct from the animals, the pinnacle of God's creation and made in God's image, the Earth and everything on it does not belong to humankind but to God; humans were to act as God's stewards and were accountable for what they did to, and with, the natural world. Indeed at times direct, indirect, option and intrinsic values were inextricably linked, e.g. during times of war the people of Israel were forbidden to cut down fruit trees to help defend themselves, even if their lives depended upon it!

A belief that biodiversity has intrinsic as well as use value is deeply rooted, even though often wilfully ignored, in many societies, cultures and faiths as well as in the Convention. This deep-seated notion of the value of biodiversity is unlikely to go away, notwithstanding its critics, and has profound implications for when we come to discuss how much biodiversity is 'enough' (see Section 4.4.2) and how we should maintain it (see Chapter 5).

4.4 How much biodiversity?

. . . as Rachel Carson reveals, human life itself may well be endangered by pesticides. Aldous Huxley, after reading Silent Spring, *remarked 'We are losing half the subject matter of English poetry'. He might well have added that we are in danger of losing the poets too.* (From the back cover of *Silent Spring* by Rachel Carson, 1968 Penguin reprint.)

4.4.1 Only one Earth

Much of the discussion thus far has emphasized the value of the components of biodiversity (biological resources), not of biodiversity itself. If all biodiversity has intrinsic value, then nothing short of all biodiversity is valuable and should be maintained. The problem arises when the intrinsic value of biodiversity conflicts with other things we value, e.g. feeding our families. So the issue becomes which value is greater. However, even from the narrow point of view that regards the elements of biodiversity merely as commodities (not that we would in any way advocate this) there is still good reason to maintain some degree of biodiversity. Plainly there are a very large number of species that have (or in the future will be found to have) known direct use value, and we may drive them extinct if we do not change our ways. This argument applies equally well to species that have indirect use value. Ultimately, the ability of Earth to sustain human life is inextricably bound up with biodiversity and as we have seen we are far from being able to create and sustain our own biosphere (see Section 4.2.2), even if such a thing was desirable. There is only one Earth, and no matter who you are biodiversity does matter. Given that biodiversity is, even from a purely selfish view, 'a good thing' and given the fact that our current practices are eroding that biodiversity (see Sections 2.3.3 & 2.5), it follows that the 'bottom line' is: how much of biodiversity can we alter, or lose, before we threaten our own existence? We are very unlikely to wipe out life in its entirety before destroying its capacity to sustain humanity.

4.4.2 How much is enough?

> *There are some marine biologists whose chief interest is in . . .*
> *rarity . . . Such collectors should to a certain extent be regarded in the same*
> *class with those philatelists who achieve a great emotional stimulation from*
> *an unusual number of perforations or a misprinted stamp. The rare animal*
> *may be of individual interest but he is unlikely to be of much consequence in*
> *any ecological picture.* Steinbeck (1990)

The question of how much biodiversity is sufficient, from an ecological point of view, has received particular attention in the context of ecosystem function. In addition to the null hypothesis of no effect, there are four ways in which ecosystem processes might respond to reductions in species richness (Lawton 1994; Johnson *et al.* 1996).

1 The diversity–stability hypothesis predicts that ecological communities will decrease in ability to recover from disturbance and in energetic efficiency (productivity) as the number of species decreases (Fig. 4.1a; Johnson *et al.* 1996).

2 The rivet hypothesis likens the species in an ecosystem to the rivets holding together an aeroplane (Fig. 4.1b; Ehrlich & Ehrlich 1981). The loss of a few rivets may go unnoticed, because they may be redundant, but beyond some threshold losses will bring about catastrophic collapse. The relationship between species richness and ecosystem function will thus be non-linear, but may follow a variety of possible trajectories.

3 The redundant species hypothesis segregates species into functional groups, with ecosystem function being little impaired by species losses if representatives of all the groups remain but being markedly impaired if all the representatives of particular groups are lost; a minimal diversity is necessary for ecosystem function, but beyond this most species are redundant in their roles (Fig. 4.1c; Walker 1992).

4 The idiosyncratic response hypothesis suggests that as diversity changes so does ecosystem function, but the magnitude and direction of change is unpredictable because individual species have complex and varied roles (Fig. 4.1d; Lawton 1994).

Which of these hypotheses best reflects reality is an important issue. Arguably it constitutes one of the most pressing questions presently facing ecologists. Experimental testing of these hypotheses, however, is not a simple matter. Whilst there have been several very important attempts, both in the laboratory and in the field (e.g. Naeem *et al.* 1994, 1995; Tilman & Downing 1994; Tilman *et al.* 1996), the interpretation, significance and generality of their results remains contentious (e.g. Johnson *et al.* 1996; Huston 1997). Particular difficulties include differentiating between cause and effect (do differences in ecosystem function result from differences in diversity or vice versa), avoiding the complications of confounding variables, scaling between systems

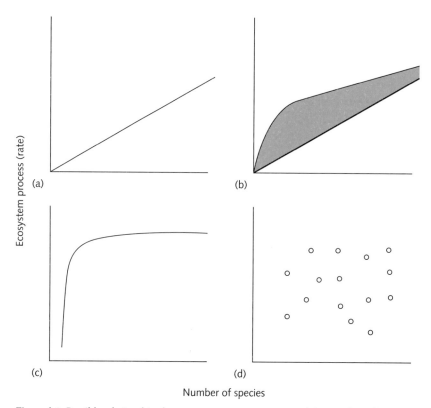

Figure 4.1 Possible relationships between ecosystem processes and the number of species in communities as predicted by (a) the diversity–stability hypothesis, (b) the rivet hypothesis, (c) the redundancy hypothesis and (d) the idiosyncratic hypothesis. (After Johnson *et al.* 1996.)

with small numbers of species that are experimentally tractable and systems with large numbers of species that are not, and adequately and appropriately replicating experimental treatments (the number of possible combinations of even small numbers of species is huge).

In view of the fundamental importance of maintaining ecosystem services

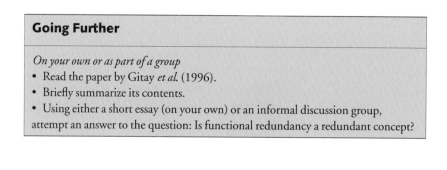

Going Further

On your own or as part of a group
• Read the paper by Gitay *et al.* (1996).
• Briefly summarize its contents.
• Using either a short essay (on your own) or an informal discussion group, attempt an answer to the question: Is functional redundancy a redundant concept?

and in the absence of any strong experimental evidence to suggest that ecosystem function is not an increasing function of diversity (and some direct support that it *is* an increasing function), it would seem foolish to do other than assume that it does so increase. The precautionary principle should apply.

4.5 Conclusion

Ultimately, of course, there is no one answer to the question of how much biodiversity is sufficient before we threaten our own existence. Rather, there are multiple answers that depend on the kind of existence we would like to enjoy. The more significant question is perhaps not how much biodiversity is sufficient, but what sort of world do we want to live in (Morowitz 1991) and what sort of world are we prepared to pay to live in? In the extreme, the distinction lies between the mentality that condemns Mozart concertos, Monet paintings or medieval cathedrals as 'functionally redundant' and that which desires to conserve such things not because they are 'useful' but because they are beautiful and because they enrich our lives (Lawton 1991).

4.6 Summary

1 Direct use values of biodiversity are concerned with the consumption or production of marketable commodities. These include food, medicine, use in biological control, industrial raw materials, recreational harvesting and ecotourism. Many present patterns of exploitation are not sustainable.

2 Indirect use values of biodiversity are more difficult to quantify, not being subject to direct trading in the marketplace, but are none the less real and important, embracing many of the services provided by biodiversity that are crucial for human well-being.

3 It is not currently possible to build artificial systems that could provide us with the life-supporting systems that natural systems provide us 'for free'.

4 Apart from present-day use value, biodiversity may be retained for the options for the use or non-use value it provides.

5 A belief that biodiversity has intrinsic value, as well as use value, is deeply rooted in many societies, cultures and faiths, and is also central to the spirit of the Convention on Biological Diversity.

6 Regardless of what different people value about biodiversity, maintaining biodiversity is essential if we are not to threaten our own existence.

7 Four hypotheses have been proposed (diversity–stability, rivet, redundant species, idiosyncratic response), in addition to the null hypothesis, to explain how ecosystem processes might respond to reduction in species richness.

8 It is difficult to answer the question: how much biodiversity is required? Perhaps the question we should be asking is: what sort of world do we want to live in?

4.7 Making connections

If biodiversity does matter, how do we go about maintaining it, particularly in the face of the present elevated levels of erosion? This issue provides the central topic for the final chapter of this book.

4.8 References

Clark, C.W. (1981) Bioeconomics. In: *Theoretical Ecology* (ed. R.M. May), pp. 387–418. Sinauer Associates, Sunderland, Massachusetts.

Cohen, J.E. & Tilman, D. (1996) Biosphere 2 and biodiversity: the lessons so far. *Science* **274**, 1150–1151.

Colwell, R.R. (1997) Microbial biodiversity and biotechnology. In: *Biodiversity II* (eds M.L. Reaka-Kudla, D.E. Wilson & E.O. Wilson), pp. 279–287. Joseph Henry Press, Washington, DC.

Crawley, M.J. (1997) Biodiversity. In: *Plant Ecology* (ed. M.J. Crawley), 2nd edn, pp. 595–632. Blackwell Science, Oxford.

Ehrlich, P.R. & Ehrlich, A.H. (1981) *Extinction: the Causes and Consequences of the Disappearance of Species.* Random House, New York.

Ehrlich, P.R. & Ehrlich, A.H. (1992) The value of biodiversity. *Ambio* **21**, 219–226.

Ehrlich, P.R. & Wilson, E.O. (1991) Biodiversity studies: science and policy. *Science* **253**, 758–762.

Eisner, T. (1989) Prospecting for nature's chemical riches. *Issues in Science and Technology* **6**, 31–36.

Filion, F.L., Foley, J.P. & Jacquemot, A.P. (1994) The economics of global ecotourism. In: *Protected Area Economics and Policy: Linking Conservation and Sustainable Development* (eds M. Munasinghe & J. McNeely), pp. 235–252. The World Bank, Washington, DC.

Gitay, H., Wilson, J.B. & Lee, W.G. (1996) Species redundancy: a redundant concept? *Journal of Ecology* **84**, 121–124.

Heywood, V.H. (ed.) (1995) *Global Biodiversity Assessment.* Cambridge University Press, Cambridge.

Huston, M.A. (1997) Hidden treatments in ecological experiments: re-evaluating the ecosystem function of biodiversity. *Oecologia* **110**, 449–460.

Johnson, K.H., Vogt, K.A., Clark, H.J., Schmitz, O.J. & Vogt, D.J. (1996) Biodiversity and the productivity and stability of ecosystems. *Trends in Ecology and Evolution* **11**, 372–377.

Krutilla, J.V. (1967) Conservation reconsidered. *American Economic Review* **57**, 778–786.

Kunin, W.E. & Lawton, J.H. (1996) Does biodiversity matter? Evaluating the case for conserving species. In: *Biodiversity: a Biology of Numbers and Difference* (ed. K.J. Gaston), pp. 283–308. Blackwell Science, Oxford.

Lande, R., Engen, S. & Saether, B.E. (1994) Optimal harvesting, economic discounting and extinction risk in fluctuating populations. *Nature* **372**, 88–90.

Lawton, J.H. (1991) Are species useful? *Oikos* **62**, 3–4.

Lawton, J.H. (1994) What do species do in ecosystems? *Oikos* **71**, 367–374.

Lean, G. & Hinrichsen, D. (eds) (1990) *Atlas of the Environment.* Helicon Publishing, Oxford.

May, R.M. (1994) The economics of extinction. *Nature* **372**, 42–43.

Morowitz, H.J. (1991) Balancing species preservation and economic considerations. *Science* **253**, 752–754.

Naeem, S., Thompson, L.J., Lawler, S.P., Lawton, J.H. & Woodfin, R.M. (1994)

Declining biodiversity can alter the performance of ecosystems. *Nature* **368**, 734–737.

Naeem, S., Thompson, L.J., Lawler, S.P., Lawton, J.H. & Woodfin, R.M. (1995) Empirical evidence that declining species-diversity may alter the performance of terrestrial ecosystems. *Philosophical Transactions of the Royal Society of London Series B* **347**, 249–262.

Norgaard, R.B. (1988) Economics of the cassava mealybug [*Phenacoccus manihoti*; Hom.: Pseudococcidae] biological control program in Africa. *Entomophaga* **33**, 3–6.

O'Toole, C. (1993) Diversity of native bees and agroecosystems. In: *Hymenoptera and Biodiversity* (eds J. LaSalle & I.D. Gauld), pp. 169–196. CAB International, Wallingford.

Perrings, C. (1995) The economic value of biodiversity. In: *Global Biodiversity Assessment* (ed. V.H. Heywood), pp. 823–914. Cambridge University Press, Cambridge.

Steinbeck, J. (1990) *Log from the Sea of Cortez*. Mandarin, London.

Thomsen, J.B., Edwards, S.R. & Mulliken, T.A. (1992) *Perceptions, Conservation and Management of Wild Birds in Trade*. TRAFFIC International, Cambridge.

Tilman, D. & Downing, J.A. (1994) Biodiversity and stability in grasslands. *Nature* **367**, 363–365.

Tilman, D., Wedin, D. & Knops, J. (1996) Productivity and sustainability influenced by biodiversity in grassland ecosystems. *Nature* **379**, 718–720.

Walker, B.H. (1992) Biodiversity and ecological redundancy. *Conservation Biology* **6**, 18–23.

Weisbrod, B. (1964) Collective consumption services of individual consumption goods. *Quarterly Journal of Economics* **77**, 71–77.

Westman, W.E. (1977) How much are nature's services worth? *Science* **197**, 960–964.

White, L. (1967) The historical roots of our ecological crisis. *Science* **155**, 1203–1207.

Wilson, E.O. (1993) Biophilia and the conservation ethic. In: *The Biophilia Hypothesis* (eds S.T. Kellert & E.O. Wilson). Island Press, Washington, DC.

World Conservation Monitoring Centre (1992) *Global Biodiversity: Status of the Earth's Living Resources*. Chapman & Hall, London.

World Resources Institute (1994) *World Resources 1994–95*. Oxford University Press, Oxford.

4.9 Further reading

Barbier, E.B., Burgess, J.C. & Folke, C. (1994) *Paradise Lost? The Ecological Economics of Biodiversity*. Earthscan, London.

Berry, R.J. (1996) Creation and the environment. *Science and Christian Belief* 7, 21–43. (*A Christian approach to the environment rediscovered?*)

Collins, S.L. & Benning, T.L. (1996) Spatial and temporal patterns in functional diversity. In: *Biodiversity: a Biology of Numbers and Difference* (ed. K.J. Gaston), pp. 253–280. Blackwell Science, Oxford. (*An excellent review of the dynamics of functional diversity.*)

Gottlieb, R.S. (ed.) (1996) *This Sacred Earth: Religion, Nature and the Environment*. Routledge, London.

Jones, C.G. & Lawton, J.H. (eds) (1995) *Linking Species and Ecosystems*. Chapman & Hall, London. (*The time has come to break down the barriers between these fields of study, and this is a major assault.*)

Kellert, S.R. (1993) Values and perceptions of invertebrates. *Conservation Biology* 7, 845–855. (*The difficulties associated with convincing people of the biodiversity value of creatures widely perceived as small and nasty.*)

Orians, G.H., Brown, G.M., Kunin, W.E. & Swierbinski, J.E. (eds) (1990) *The Preservation and Valuation of Biological Resources.* University of Washington Press, Seattle. (*Good on valuation of biodiversity, including genetic resources.*)

Pearce, D.W. & Moran, D. (1994) *The Economic Value of Biological Diversity.* Earthscan, London.

Peters, C., Gentry, A. & Mendelsohn, R. (1989) Valuation of an Amazonian rainforest. *Nature* **339**, 655–656.

Reid, W.V. & Miller, K.R. (1989) *Keeping Options Alive.* World Resources Institute, Washington, DC.

Reid, W.V., Laird, S.A., Meyer, C.A., Gamez, R., Sittenfeld & A., Janzen *et al.* (1993) *Biodiversity Prospecting: Using Genetic Resources for Sustainable Development.* World Resources Institute, Washington, DC. (*An extremely important subject that we hardly touch on here; the list of biotechnology options is as long as the string of coauthors.*)

Schulze, E-D. & Mooney, H.A. (eds) (1993) *Biodiversity and Ecosystem Function.* Springer Verlag, Berlin. (*A landmark volume in this area, but already being overtaken by events?*)

Wilson, E.O. & Peter, F.M. (eds) (1988) *BioDiversity.* National Academy Press, Washington, DC. (*Many contributions address issues of use and value.*)

5 Maintaining biodiversity

5.1 Introduction

Concern over changing levels of, and patterns in, biological diversity is not new. Over a century ago Alfred Russel Wallace, who independently of Charles Darwin proposed the theory of natural selection, wrote:

> *We live in a zoologically impoverished world, from which all the hugest, and fiercest, and strangest forms have recently disappeared . . . yet it is surely a marvellous fact, and one that has hardly been sufficiently dwelt upon, this sudden dying out of so many large Mammalia, not in one place only but over half the land surface of the globe.* Wallace (1876)

Nevertheless, there has never been a time like the present when the issues surrounding biodiversity have been more pressing, not just nationally but also globally.

Use of the term 'biodiversity' arose in the context of, and has remained firmly wedded to, concerns over the loss of the natural environment and its contents. The importance of this connection cannot be overstated. In defining biodiversity in this book, we have relied heavily on the Convention on Biological Diversity (see Section 1.2). This was not solely a matter of convenience (though it does sidestep the often thorny debates on 'What is biodiversity?'). It underscores our belief that, for better or worse, this is perhaps the single most important document for the maintenance of biodiversity (see Section 5.3). Having examined briefly the main features and patterns of, and the value placed on, biodiversity (Chapters 2, 3 & 4), what better way to discuss its maintenance than to go back to the articles contained in the Convention.

Before discussing how to maintain biodiversity, in the context of the impact humankind has had, and is having, we must first try to appreciate the scale of the human enterprise. This chapter commences by presenting some basic facts that assist us to do just that. Some background to the Convention on Biological Diversity is then given as a prelude to a discussion of the articles that directly impact on biodiversity, as covered in Chapters 1–4. Each of the articles chosen is reproduced in full, followed by some commentary. We would encourage readers not to be deterred by the legal language (with its multiple caveats and subclauses) of the sections of the Convention that are quoted; the caveats, etc. were necessary to achieve a document which so many countries could sign up to. Although at times rather formidable, the underlying ideas remain simple to understand and are amplified in the accompanying text.

5.2 Scale of the human enterprise

Even to begin to comprehend the impact of human activity on biodiversity, we need to try to understand something of the scale of the human enterprise. A few observations help us to do this.

1 Population growth: the world's human population is estimated to total about 5.7 billion and to be increasing by more than 86 million annually (World Resources Institute 1996). Population growth has been slow for most of human existence but over the past 200 years the rate has increased dramatically (Fig. 5.1).

2 Urbanization: within the next decade more than half of the world's human population will be living in urban areas. Cities are already reaching unprecedented sizes (Tokyo, 27 million; São Paulo, 16.4 million; Bombay, 15 million), and by 1990 there were 21 megacities (population exceeding 8 million; World Resources Institute 1996).

3 Energy use: Ehrlich (1995) estimates that from before the agricultural revolution to the present time, total power consumption by humanity multiplied roughly 7000–13000 fold, from 0.001–0.002 TW $(1\,TW = 10^{12}\,W)$ to 13 TW. Global commercial energy production in 1993 reached 338 EJ $(1\,EJ =$

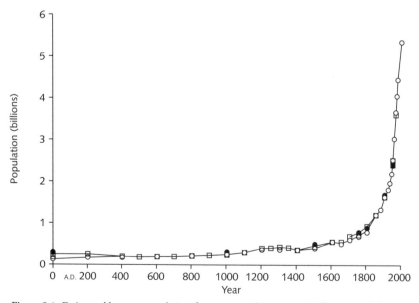

Figure 5.1 Estimated human population from AD 1 to the present. Different symbols represent estimates from different sources. (After Cohen 1996.)

Table 5.1 The pattern of human disturbance amongst biomes. Undisturbed areas have a record of primary vegetation and no evidence of disturbance, combined with a very low human population density. Partially disturbed areas have a record of shifting or extensive agriculture, evidence of secondary vegetation, livestock over carrying capacity or other evidence of human disturbance. Human dominated areas have a record of permanent agriculture or urban settlement, removal of primary vegetation or record of desertification or other permanent degradation. (After Hannah *et al.* 1995.)

Biome	Total area (km²)	Area undisturbed (%)	Area partially disturbed (%)	Area human dominated (%)
Temperate broadleaf forests	9 519 442	6.1	12.0	81.9
Evergreen sclerophyllous forests	6 559 728	6.4	25.8	67.8
Temperate grasslands	12 074 494	27.6	32.0	40.4
Subtropical and temperate rainforests	4 232 299	33.0	20.9	46.1
Tropical dry forests	19 456 659	30.5	41.1	45.9
Mixed mountain systems	12 133 746	29.3	45.0	25.6
Mixed island systems	3 256 096	46.6	11.6	41.8
Cold deserts/semi-deserts	10 930 762	45.4	46.1	8.5
Warm deserts/semi-deserts	29 242 021	55.8	32.0	12.2
Tropical humid forests	11 812 012	63.2	11.9	24.9
Tropical grasslands	4 797 090	74.0	21.3	4.7
Temperate needleleaf forests	18 830 709	81.7	6.4	11.8
Tundra and Arctic desert	20 637 953	99.3	0.7	0.3

10^{18} J, or about 163 million barrels of oil), 40% greater than in 1973. Total energy consumption rose to 326 EJ, 49% greater than 20 years before (World Resources Institute 1996).

4 Emissions: global emissions of carbon dioxide in 1992 amounted to 26.4 Gt ($1 Gt = 10^9$ t), of which 84% was from industrial activity (World Resources Institute 1996).

5 Primary productivity: humans use, co-opt or destroy approximately 40% of all potential terrestrial net primary productivity (Vitousek *et al.* 1986).

6 Land-use: world changes in land-use over the past three centuries (i.e. since 1700) have been estimated to include a 19% reduction in the extent of forests and woodlands, an 8% reduction in grasslands and pasture, and a 466% increase in croplands brought into cultivation (Richards 1990).

7 Habitat disturbance: human disturbance is evident in every biome on Earth, being most marked in temperate broadleaf and evergreen sclerophyllous forests (< 6.5% undisturbed; Table 5.1).

8 Land degradation: approximately 43% of the Earth's terrestrial vegetated

Table 5.2 Factors affecting current levels of biodiversity. (After McNeely *et al.* 1995.)

Immediate causes
Exploitation of wild living resources
Expansion of agriculture, forestry and aquaculture
Habitat loss and fragmentation
Species introductions
Pollution of soil, water and atmosphere
Global climate change

Underlying causes
Human social organization
Growth of human population
Patterns of natural resource consumption
Global trade
Economic systems and policies that fail to value the environment and its resources
Inequity in ownership, management and flow of benefits from use and conservation of biological resources

---

surface is estimated to have a diminished capacity to supply benefits to humanity because of recent, direct impacts of land-use (Daily 1995).

It is inconceivable that an enterprise of this scale would not have major detrimental impacts on biodiversity. The high levels of global and local extinction (see Section 2.5) testify to this fact. A number of human factors can be recognized that derive directly or indirectly from the above and that contribute to the erosion of biological diversity (Table 5.2). Decline in biodiversity includes all those changes that are associated with reducing or simplifying biological heterogeneity, from individuals to regions (Walker 1992).

5.3 The Convention on Biological Diversity

At the heart of attempts to address the detrimental impacts of human activity on biodiversity lies the framework provided by the Convention on Biological Diversity. The Convention constitutes a historic commitment by nations of the world (though sadly not all of them have yet ratified or even signed). It is the first time that biodiversity is comprehensively addressed in a binding global treaty, the first time that genetic diversity is specifically covered and the first time that the conservation of biodiversity is recognized as the common concern of humankind (Glowka *et al.* 1994).

The Convention comprises 42 articles (Table 5.3), concerning issues that include its objectives, the practical obligations of each signatory, the policies to be followed and the use of terms. Below we discuss a selection of these articles and their relationship to the maintenance of biodiversity. As should rapidly become apparent, the maintenance of biodiversity is much more than a matter

Table 5.3 The 42 articles of the Convention on Biological Diversity.

1 Objective
2 Use of Terms
3 Principle
4 Jurisdictional Scope
5 Cooperation
6 General Measures for Conservation and Sustainable Use
7 Identification and Monitoring
8 *In-Situ* Conservation
9 *Ex-Situ* Conservation
10 Sustainable Use of Components of Biological Diversity
11 Incentive Measures
12 Research and Training
13 Public Education and Awareness
14 Impact Assessment and Minimizing Adverse Impacts
15 Access to Genetic Resources
16 Access to and Transfer of Technology
17 Exchange of Information
18 Technical and Scientific Cooperation
19 Handling of Biotechnology and Distribution of its Benefits
20 Financial Resources
21 Financial Mechanism
22 Relationship with other International Conventions
23 Conference of the Parties
24 Secretariat
25 Subsidiary Body on Scientific, Technical and Technological Advice
26 Reports
27 Settlement of Disputes
28 Adoption of Protocols
29 Amendment of the Convention or Protocols
30 Adoption and Amendment of Annexes
31 Right to Vote
32 Relationship between this Convention and its Protocols
33 Signature
34 Ratification, Acceptance or Approval
35 Accession
36 Entry into Force
37 Reservations
38 Withdrawals
39 Financial Interim Arrangements
40 Secretariat Interim Arrangements
41 Depositary
42 Authentic Texts

of nature reserves and keeping the most highly threatened species from extinction.

5.3.1 Article 1. Objectives of the Convention

The objectives of the Convention are threefold:

> *The conservation of biological diversity, the sustainable use of its components, and the fair and equitable sharing of the benefits arising from the utilization of genetic resources.*

This is the heart of the Convention, establishing the framework and context for the subsequent articles, and its overall sense of direction.

Right at the outset the Convention recognizes some of the main strands that must be involved in the future interaction of humanity with biodiversity. Diversity must be maintained, if only because failure to do so would be to imperil human existence (cf. Section 4.4.1). This can only be achieved through sustainable use and only if the benefits arising from the use are fairly and equitably distributed. This reflects a general acceptance that there are social contexts to conservation actions.

> *To avoid possible confusion, 'sustainable use' is defined (in Article 2) as: the use of components of biological diversity in a way and at a rate that does not lead to the long term decline of biological diversity, thereby maintaining its potential to meet the needs and aspirations of present and future generations.*

5.3.2 Article 6. General Measures for Conservation and Sustainable Use

This is perhaps one of the most far-reaching and significant articles in the Convention, and reads as follows:

> *Each Contracting Party shall, in accordance with its particular conditions and capabilities:*
>
> *(a) Develop national strategies, plans or programmes for the conservation and sustainable use of biological diversity or adapt for this purpose existing strategies, plans or programmes which shall reflect,* inter alia, *the measures set out in this Convention relevant to the Contracting Party concerned; and*
>
> *(b) Integrate, as far as possible and as appropriate, the conservation and sustainable use of biological diversity into relevant sectoral or cross-sectoral plans, programmes and policies.*

In short, the conservation and sustainable use of biodiversity are not expected to emerge fortuitously in each nation. Indeed they will not, as the recent history of biodiversity testifies (see Sections 2.3.3, 2.5 & 2.6). Rather, the Convention obliges nations to establish mechanisms for bringing these things about, or for developing these mechanisms if they already exist. Strategies, plans and programmes can be seen as a chronological series of steps whereby specific recommendations are turned into methods of achieving those ends and thence into action on the ground (Glowka *et al.* 1994). They will inevitably have to be dynamic, under continual refinement and development, in order to respond to the changing circumstances of biodiversity in a particular nation. If they are to be effective, these national strategies, plans and programmes will not be easy to formulate, as they will have to touch on multiple (perhaps even most) human activities. They will thus have to be integrated with policies in fields as diverse as agriculture, education, employment, energy, health, industry and transport. *If they are to be truly effective, the strategies, plans and programmes for conserving and sustainably using a nation's biological diversity will have to become central to the way in which that nation's affairs are run.*

5.3.3 Article 7. Identification and Monitoring

In order to know whether strategies, programmes and plans for conservation and sustainable use are appropriate and are working effectively, it will be necessary to gather suitable information. Article 7 places such an obligation on signatories to the Convention (Annex I is given in Table 5.4):

> *Each Contracting Party shall, as far as possible and as appropriate, in particular for the purposes of Articles 8–10:*
> *(a) Identify components of biological diversity important for its conservation and sustainable use having regard to the indicative list of categories set down in Annex I;*
> *(b) Monitor, through sampling and other techniques, the components of biological diversity identified pursuant to subparagraph (a) above, paying particular attention to those requiring urgent conservation measures and those which offer the greatest potential for sustainable use;*

Table 5.4 Annex I of the Convention on Biological Diversity: Identification and Monitoring.

1 Ecosystems and habitats: containing high diversity, large numbers of endemic or threatened species, or wilderness; required by migratory species; of social, economic, cultural or scientific importance; or, which are representative, unique or associated with key evolutionary or other biological processes;
2 Species and communities which are: threatened; wild relatives of domesticated or cultivated species; of medicinal, agriculture or other economic value; or social, scientific or cultural importance; or importance for research into the conservation and sustainable use of biological diversity, such as indicator species; and
3 Described genomes and genes of social, scientific or economic importance.

*(c) Identify processes and categories of activities which have or are likely to
have significant adverse impacts on the conservation and sustainable use of
biological diversity, and monitor their effects through sampling and other
techniques; and*

*(d) Maintain and organize, by any mechanism data, derived from
identification and monitoring activities pursuant to subparagraphs (a), (b)
and (c) above.*

The combination of the paucity of knowledge of biodiversity and its great
magnitude (see Chapters 2 & 3) makes it impossible to identify or monitor all
of the components of biodiversity that lie within a nation's borders. Article 7
and its associated Annex I therefore concentrate these undertakings in two
directions: first, on those components that are considered to be important for
the conservation and sustainable use of biodiversity; and, second, on those
activities likely to have the most substantial impacts on this conservation and
use. Much of this will require the acquisition of entirely new information,
while it will be possible to use some existing data (see Chapters 1, 2 & 3),
perhaps freshly collated. Combined, this will have benefits far beyond the
Convention, serving to improve overall understanding of biodiversity. This
will be facilitated by the final clause of this article.

The ease with which nations can begin to fulfil the requirements of this
article will vary dramatically, on the basis of existing knowledge alone (cf. final
comments on Article 8).

Going Further

As a group

• Choose an area that is in close proximity to where you are and/or one that you
can easily acquire relevant information on (from a library, for instance). The area
could be a local park, wood, forest, a piece of coastline, a wilderness area or even an
entire region or country – whatever you think. If it is possible for you to arrange a
visit to the site, all the better.

• Imagine that you and your colleagues are part of a working party called together
to implement Article 7c of the Convention. Your remit is to 'identify processes and
categories of activities which have or are likely to have significant adverse impacts
on the conservation and sustainable use of biological diversity' *in your chosen area.*
From relevant literature, and from your own observations if possible, put together a
poster that does exactly this. Remember the poster should introduce the site itself
as well as identify areas for concern.

5.3.4 Article 8. *In-situ* Conservation

Article 8 embodies the principal obligations for the conservation of biological
diversity. Although it is one of the longer articles in the Convention, and thus

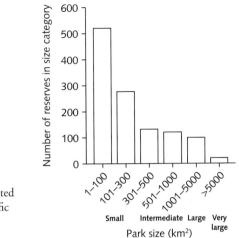

Figure 5.2 Size distribution of protected areas in 23 countries of the Indo-Pacific region. (After Dinerstein & Wikramanayake 1993.)

may appear especially daunting, it is so important that we must consider all of it. However, to make the task a little less onerous we will divide it into manageable sections.

> *Each Contracting Party shall, as far as possible and appropriate:*
> *(a) Establish a system of protected areas or areas where special measures need to be taken to conserve biological diversity;*
> *(b) Develop, where necessary, guidelines for the selection, establishment and management of protected areas or areas where special measures need to be taken to conserve biological diversity;*

Protected area systems or networks are required to be established as a central plank of a national strategy for conserving biodiversity. Nearly 10 000 protected areas, spread amongst virtually all countries in the world, are recognized by the IUCN Commission on Parks and Protected Areas, although the objectives for their designation and their effectiveness in conserving biodiversity are varied. Most such areas are small (Fig. 5.2). In few, if any, countries have sets of protected areas been designated on the basis of an objective procedure aimed at conserving maximal biodiversity, although techniques are currently available to assist in such a process (e.g. Pressey *et al.* 1993; Vane-Wright 1996).

> *(c) Regulate or manage biological resources important for the conservation of biological diversity whether within or outside protected areas, with a view to assuring their conservation and sustainable use;*
> *(d) Promote the protection of ecosystems, natural habitats and the maintenance of viable populations of species in natural surroundings;*
> *(e) Promote environmentally sound and sustainable development in areas adjacent to protected areas with a view to furthering protection of these areas;*

Systems of protected areas alone will not be adequate for the purpose of conserving biodiversity, if for no other reason than that they do not exist in isolation from the areas around them. What happens in the broader environment, in which such protected areas are embedded, must inevitably impinge upon them. These paragraphs of Article 8 therefore require the management of biological resources both within protected areas and outside of them (i.e. the general protection of ecosystems and populations wherever they occur), and so ensure that development in areas adjacent to protected areas does not undermine the capacity of those protected areas to conserve biodiversity.

> (f) *Rehabilitate and restore degraded ecosystems and promote the recovery of threatened species,* inter alia, *through the development and implementation of plans or other management strategies;*

The conservation of biodiversity is not simply about maintaining things the way they presently are. As we have seen, few (if any) areas are pristine and untouched, directly or indirectly, by human hand and many are severely degraded (see Table 5.1). A creative approach to restoration is thus also required and has given rise to the emergence of the science of restoration ecology (Jordan *et al.* 1990; Pywell & Putwain 1996).

> (g) *Establish or maintain means to regulate, manage or control the risks associated with the use and release of living modified organisms resulting from biotechnology which are likely to have adverse environmental impacts that could affect the conservation and sustainable use of biological diversity, taking also into account the risks to human health;*
> (h) *Prevent the introduction, control or eradicate those alien species which threaten ecosystems, habitats or species;*

The impacts on biodiversity associated with the introduction of alien species have already been mentioned (see Section 2.6); plainly, actions to ameliorate these effects are a necessary part of an effective conservation strategy. The need to combat the possible risks associated with the use and release of living 'modified' organisms (which include genetically modified organisms) is particularly highlighted.

> (i) *Endeavour to provide the conditions needed for compatibility between present uses and the conservation of biological diversity and the sustainable use of its components;*
> (j) *Subject to its national legislation, respect, preserve and maintain knowledge, innovations and practices of indigenous and local communities embodying traditional lifestyles relevant for the conservation and sustainable use of biological diversity and promote their wider application with the approval and involvement of the holders of such knowledge, innovations and practices and encourage the equitable*

sharing of the benefits arising from the utilization of such knowledge,
innovations and practices;

Intuitively, support for the conservation of biological diversity will be less when necessary changes conflict with present uses (see Section 4.2). The first of these paragraphs requests that parties to the Convention should minimize these conflicts, although plainly this will often be difficult and, at times, impossible. The second paragraph recognizes that the knowledge, innovations and practices of indigenous and local communities may be pertinent to the conservation and sustainable use of biodiversity, and that this cultural relevance should be promoted, to the benefit of its custodians.

(k) Develop or maintain necessary legislation and/or other
regulatory provisions for the protection of threatened species and populations;
(l) Where a significant adverse effect on biological diversity has been
determined pursuant to Article 7, regulate or manage the relevant processes
and categories of activities; and
(m) Cooperate in providing financial and other support for in-situ
conservation outlined in subparagraphs (a) to (l), particularly to developing
countries.

These paragraphs all concern mechanisms for conserving biodiversity, including the development of appropriate legislation, the regulation and management of processes and activities that from the gathering of suitable information (as outlined in Article 7) have been found to be detrimental to biodiversity, and the provision of financial and other support to developing countries. The final paragraph reflects a recurrent theme of the Convention by recognizing that the resources available for the conservation and sustainable use of biodiversity are not evenly distributed, and that the poorer countries (where arguably most biodiversity is found; see Section 3.4.1) will require support from the richer if these ends are to be achieved. In addition, it should be remembered, the relative impacts of factors affecting biodiversity are not the same in poorer and richer countries (Fig. 5.3).

5.3.5 Article 9. *Ex-situ* Conservation

Conservation actions have traditionally been divided into *in situ* and *ex situ*; having dealt with the former in Article 8, the Convention moves on to the latter in Article 9.

Each Contracting Party shall, as far as possible and as appropriate, and
predominantly for the purpose of complementing in-situ measures:
(a) Adopt measures for the ex-situ conservation of components of biological
diversity, preferably in the country of origin of such components;
(b) Establish and maintain facilities for ex-situ conservation of and research

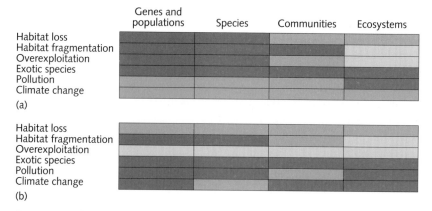

Figure 5.3 Relative impacts of factors affecting terrestrial biotic diversity in (a) poor and (b) rich countries. Shading indicates intensity of impact, from black (highest) to light grey (lightest). (After Soulé 1991.)

on plants, animals and microorganisms, preferably in the country of origin of genetic resources;

(c) Adopt measures for the recovery and rehabilitation of threatened species and for their reintroduction into their natural habitats under appropriate conditions;

(d) Regulate and manage collection of biological resources from natural habitats for ex-situ *conservation purposes so as not to threaten ecosystems and* in-situ *populations of species, except where special temporary* ex-situ *measures are required under subparagraph (c) above; and*

(e) Cooperate in providing financial and other support for ex-situ *conservation outlined in subparagraphs (a) to (d) above and in the establishment and maintenance of* ex-situ *conservation facilities in developing countries.*

Ex-situ conservation measures may include seed banks, sperm and ova banks, culture collections (e.g. of plant tissues), captive breeding of animals, and artificial propagation of plants. The relative costs and benefits of *ex-situ* conservation have been much debated (e.g. Tudge 1992; Rahbek 1993; Hurka 1994; Balmford *et al.* 1995; Frankel *et al.* 1995). Clearly it will play only a very secondary role to *in-situ* conservation, as implied by the opening statement of this article.

> **Going Further**
>
> *On your own*
> • Draw up a list of the pros and cons of the *ex-situ* conservation of plants and animals.
> • If you were asked what you thought of the use of zoos in *ex-situ* conservation what would your considered answer be and why?
> *Or as a group*
> • Debate the following question: *Ex-situ* conservation – is this the road we want to go down?
> • Split into two groups (three to five persons). One group will argue 'yes' to the question and the other will argue 'no'.
> • In your groups, read the recommended literature and decide on what you are going to present and how.
> • Each group should have no more than 10 minutes to present their case. After this time the groups should be allowed to cross-examine each other but for no more than 20 minutes.
> • Finally, time should be allocated at the end (approximately 15 minutes) when group members can express their own personal opinions and see if any consensus emerges.
>
> **References**
> Balmford *et al.* (1995, 1996), Henry (1997)

5.3.6 Article 10. Sustainable Use of Components of Biological Diversity

The sustainable use of biological diversity is one of the objectives of the Convention (Article 1). Article 10 embodies the obligations for attaining this goal.

> *Each Contracting Party shall, as far as possible and as appropriate:*
> *(a) Integrate consideration of the conservation and sustainable use of biological resources into national decision-making;*
> *(b) Adopt measures relating to the use of biological resources to avoid or minimize adverse impacts on biological diversity;*
> *(c) Protect and encourage customary use of biological resources in accordance with traditional cultural practices that are compatible with conservation or sustainable use requirements;*
> *(d) Support local populations to develop and implement remedial action in degraded areas where biological diversity has been reduced; and*
> *(e) Encourage cooperation between its governmental authorities and its private sector in developing methods for sustainable uses of biological resources.*

In essence, sustainable use is to be attained by its integration into national planning, by minimizing the adverse impacts of use on biodiversity. In Section

4.2 we highlighted that biodiversity is presently being used in self-evidently non-sustainable ways (management approaches have often focused on maximizing short-term yield and economic gain rather than long-term sustainability). Attaining the goals of this article will therefore necessitate a major shift in the present patterns of exploitation of biological resources.

Sustainable use requires the support of local peoples, and the protection and encouragement of customary use is one way in which to achieve this. However, it is important to distinguish those traditional uses that are compatible with conservation and sustainable use from those which are not. For example, the widespread belief that 'primitive' peoples have no appreciable adverse impact on their environment is, expressed in such a generic fashion, simply a myth (Milberg & Tyrberg 1993). The impact of prehistoric peoples on island faunas (see Section 2.5.1) provides ample evidence that this is not the case.

5.3.7 Article 11. Incentive Measures

Biodiversity loss is driven in major part by economic forces (see Table 5.1).

Table 5.5 The goal, principles and objectives of the UK Action Plan. (From Anon. 1994.)

Overall goal
To conserve and enhance biological diversity within the UK and to contribute to the conservation of global biodiversity through all appropriate mechanisms.

Underlying principles
1 Where biological resources are used, such use should be sustainable.
2 Wise use should be ensured for non-renewable resources.
3 The conservation of biodiversity requires the care and involvement of individuals and communities as well as Governmental processes.
4 Conservation of biodiversity should be an integral part of Government programmes, policy and action.
5 Conservation practice and policy should be based upon a sound knowledge base.
6 The precautionary principle should guide decisions.

Objectives for conserving biodiversity
1 To conserve and where practicable to enhance:
   (a) the overall populations and natural ranges of native species and the quality and range of wildlife habitats and ecosystems;
   (b) internationally important and threatened species, habitats and ecosystems;
   (c) species, habitats and natural and managed ecosystems that are characteristic of local areas;
   (d) the biodiversity of natural and semi-natural habitats where this has been diminished over recent past decades.
2 To increase public awareness, and involvement in, conserving biodiversity.
3 To contribute to the conservation of biodiversity on a European and global scale.

Article 11 is an attempt to harness these same forces to its conservation and sustainability.

> *Each Contracting Party shall, as far as possible and as appropriate,*
> *adopt economically and socially sound measures that act as incentives*
> *for the conservation and sustainable use of components of biological*
> *diversity.*

Put simply, the obligation is to adopt measures that encourage conservation and sustainable use (Glowka *et al.* 1994).

5.4   Responses to the Convention

In accordance with Article 6 of the Convention, a number of countries, including Canada, Chile, Germany, Indonesia, Norway, the UK and Vietnam are beginning to develop, or have even published, national biodiversity strategies (general policy instruments to identify strategic needs) or action plans (practical documents that identify what is to be done and who is to do what) (Miller *et al.* 1995). For example, publication of the UK Action Plan (Anon. 1994) represents such a direct governmental response to Article 6. Its goal, principles and objectives are listed in Table 5.5. It remains to be seen how many countries, including those that have ratified or only signed and those that have not yet signed, follow through on the guidelines laid down in the Convention and the extent to which those commitments are in practice acted upon.

5.5   Conclusions

> *. . . there is unfortunately no precedent for 5 billion human beings suddenly*
> *sharing an enlightened vision of the future.* Flesness (1992)

Throughout this book there is one clear central message. Biodiversity is not simply a topic of esoteric or academic interest; rather in a multiplicity of ways it impinges on peoples' lives. The Convention on Biological Diversity is a bold attempt to ensure that the interaction is a healthy one. It is a far from perfect document, and has been criticized variously for the relative weighting it gives to different issues, for its omissions and for its caveats and equivocation. It is, in this sense, undoubtedly also a very human document. This highlights perhaps the most important point of all. Most of the loss of biodiversity, with all its contingent consequences, results from the independent decisions of many millions of individual users of environmental resources. All of us.

Silent Spring, by Rachel Carson, first published in the early 1960s, is widely considered to have been a milestone in bringing to public awareness the colossal impact of humankind (mainly in the form of pesticide use) on biodiversity. She dedicated the book to Albert Schweitzer whom she quotes as saying 'Man has lost the capacity to foresee and to forestall. He will end by destroying the earth'. Thankfully, not only did *Silent Spring* inform but it also to some extent changed attitudes towards the indiscriminate use of pesticides and abuse of the environment more generally. The imminence of a world where 'no birds sing' was postponed, if not averted. Arguably, public perception of environmental or green issues and awareness of the importance of our forests, our animals and life on our planet generally has grown, and is much greater now than it was when Carson wrote her book. Admittedly the political and social will for change varies from country to country. It can be slow and is often lacking, but clearly it is there. Concern for 'our environment' or 'the natural environment' is evidenced by the immense popularity of wildlife programmes and books, the 'recycling culture' and the requirement for impact assessment prior to development that is emerging in some nations, by the fact that environmental issues are increasingly seen by politicians as having importance, and by the large number of high-profile scientific conferences and conventions that highlight current ecological problems. It is easy to be cynical about some, or all, of these 'hopeful signs', but it is difficult to see that we can afford the luxury of such cynicism given the present biodiversity crisis.

The Convention on Biological Diversity is currently one of our 'best hopes', presenting what we *do* foresee and proposing practical ways in which we *can* forestall. The words of Schweitzer that front *Silent Spring* do not have to form the concluding paragraph of this text; indeed we wonder if *any* concluding paragraph attempted might look out of place or be premature. Even in the time taken to write this book, new and important ideas have been floated, novel data have been collected and collated, and new and crucial experiments have been reported. Understanding of biodiversity, and what is done with that understanding, is still being shaped. Biodiversity is a 'hot topic' that, as should be apparent from all of the preceding chapters, cannot yet be tied up as a nice complete bundle and delivered neatly and tidily to the *fait accompli* box. Therefore we shall exit not with 'The End' (with all its various connotations) but, perhaps more appropriately, with an exhortation to 'watch this space'!

5.6 Summary

1 The impact of human activity on biodiversity, via population growth, urbanization, energy, land-use, pollution and habitat destruction is huge and it is global.
2 The Convention on Biological Diversity is one of the main global attempts to set an agenda for maintaining biodiversity.
3 The main objectives of the Convention are the conservation of biological diversity, the sustainable use of its components, and the fair and equitable sharing of the benefits arising from the utilization of genetic resources.
4 The articles of the Convention discussed in this chapter are those concerned with identifying, monitoring and conserving biodiversity, together with measures to ensure sustainable use of the Earth's resources.

5   A number of countries have begun to develop and publish action plans in accordance with the articles of the Convention.

5.7 References

Anon. (1994) *Biodiversity: the UK Action Plan.* HMSO, London.

Balmford, A., Leader-Williams, N. & Green, M.J.B. (1995) Parks or arks: where to conserve threatened mammals? *Biodiversity and Conservation* 4, 595–607.

Balmford, A., Mace, G.M. & Leader-Williams, N. (1996) Designing the ark: setting priorities for captive breeding. *Conservation Biology* 10, 719–727.

Cohen, J.E. (1996) *How Many People Can the Earth Support?* Norton, New York.

Daily, G.C. (1995) Restoring value to the World's degraded lands. *Science* 269, 350–354.

Dinerstein, E. & Wikramanayake, E.D. (1993) Beyond 'hotspots': how to prioritize investments to conserve biodiversity in the Indo-Pacific region. *Conservation Biology* 7, 53–65.

Ehrlich, P.R. (1995) The scale of the human enterprise and biodiversity loss. In: *Extinction Rates* (eds J.H. Lawton & R.M. May), pp. 214–226. Oxford University Press, Oxford.

Flesness, N.R. (1992) Living collections and biodiversity. In: *Systematics, Ecology and the Biodiversity Crisis* (ed. N. Eldredge), pp. 178–184. Columbia University Press, New York.

Frankel, O.H., Brown, A.H.D. & Burdon, J.J. (1995) *The Conservation of Plant Biodiversity.* Cambridge University Press, Cambridge.

Glowka, L., Burhenne-Guilmin, F., Synge, H., McNeely, J.A. & Gündling, L. (1994) *A Guide to the Convention on Biological Diversity.* IUCN, Gland & Cambridge.

Hannah, L., Carr, J.L. & Lankerani, A. (1995) Human disturbance and natural habitat: a biome level analysis of a global data set. *Biodiversity and Conservation* 4, 128–155.

Henry, J.-P. (1997) Integrating *in situ* and *ex situ* conservation. *Plant Talk* (January), 23–25.

Hurka, H. (1994) Conservation genetics and the role of botanical gardens. In: *Conservation Genetics* (eds V. Loeschcke, J. Tomiuk & K. Jain), pp. 371–380. Birkhauser Verlag, Basel.

Jordan III, W.R., Gilpin, M.E. & Aber, J.D. (eds) (1990) *Restoration Ecology: a Synthetic Approach to Ecological Research.* Cambridge University Press, Cambridge.

McNeely, J.A., Gadgil, M., Leveque, C., Padoch, C. & Redford, K. (1995) Human influences on biodiversity. In: *Global Biodiversity Assessment* (ed. V.H. Heywood), pp. 711–821. Cambridge University Press, Cambridge.

Milberg, P. & Tyrberg, T. (1993) Naïve birds and noble savages: a review of man-caused prehistoric extinctions of island birds. *Ecography* 16, 229–250.

Miller, K., Allegretti, M.H., Johnson, N. & Jonsson, B. (1995) Measurement for conservation of biodiversity and sustainable use of its components. In: *Global Biodiversity Assessment* (ed. V.H. Heywood), pp. 915–1061. Cambridge University Press, Cambridge.

Pressey, R.L., Humphries, C.J., Margules, C.R., Vane-Wright, R.I. & Williams, P.H. (1993) Beyond opportunism: key principles for systematic reserve selection. *Trends in Ecology and Evolution* 8, 124–128.

Pywell, R. & Putwain, P. (1996) Restoration and conservation gain. In: *Conservation Biology* (ed. I.F. Spellerberg), pp. 203–221. Longman, Harlow.

Rahbek, C. (1993) Captive breeding: a useful tool in the preservation of biodiversity? *Biodiversity and Conservation* 2, 426–437.

Richards, J.F. (1990) Land transformation. In: *The Earth as Transformed by Human Action* (eds B.L. Turner, W.C. Clark, R.W. Kates, J.F. Richards, J.T. Mathews & W.B. Meyer), pp. 163–178. Cambridge University Press, New York.

Soulé, M.E. (1991) Conservation: tactics for a constant crisis. *Science* 253, 744–749.

Tudge, C. (1992) *Last Animals at the Zoo. How Mass Extinction Can Be Stopped.* Oxford University Press, Oxford.

Vane-Wright, R.I. (1996) Identifying priorities for the conservation of biodiversity: systematic biological criteria within a socio-political framework. In: *Biodiversity: a Biology of Numbers and Difference* (ed. K.J. Gaston), pp. 309–344. Blackwell Science, Oxford.

Vitousek, P.M., Ehrlich, P.R., Ehrlich, A.H. & Matson, P.A. (1986) Human appropriation of the products of photosynthesis. *BioScience* 36, 368–373.

Walker, B.H. (1992) Biodiversity and ecological redundancy. *Conservation Biology* 6, 18–23.

Wallace, A.R. (1876) *The Geographical Distribution of Animals with the Study of the Relations of Living and Extinct Faunas as Elucidating the Past Changes of the Earth's Surface.* Harper, New York.

World Resources Institute (1996) *World Resources 1996–97.* Oxford University Press, Oxford.

5.8 Further reading

5.8.1 The Convention on Biological Diversity

Johnson, S.P. (1993) *The Earth Summit: The United Nations Conference on Environment and Development (UNCED).* Graham & Trotman, London.

McConnell, F. (1996) *The Biodiversity Convention: a Negotiating History.* Kluwer Law International, London. (*A fascinating account of negotiating the Convention by the head of the UK delegation.*)

World Conservation Monitoring Centre (1992) *Global Biodiversity: Status of the Earth's Living Resources.* Chapman & Hall, London.

The Convention on Biological Diversity and all of the material associated with it is accessible at http://www.unep.ch/biodiv.html.

5.8.2 Conserving biodiversity

Forey, P.L., Humphries, C.J. & Vane-Wright, R.I. (eds) (1994) *Systematics and Conservation Evaluation.* Clarendon Press, Oxford. (*An important treatise on methodology, mostly still state of the art.*)

McNeely, J.A., Miller, K.R., Reid, W.V., Mittermeier, R.A. & Werner, T.B. (1990) *Conserving the World's Biological Diversity.* IUCN, Gland; WRI, CI, WWF-US and the World Bank, Washington, DC.

Miller, K., Allegretti, M.H., Johnson, N. & Jonsson, B. (1995) Measures for conservation of biodiversity and sustainable use of its components. In: *Global Biodiversity Assessment* (ed. V.H. Heywood), pp. 915–1061. Cambridge University Press, Cambridge.

WRI/IUCN/UNEP (1992) *Global Biodiversity Strategy: Guidelines for Action to Save, Study, and Use Earth's Biotic Wealth Sustainably and Equitably.* World Resources Institute, Washington, DC; World Conservation Union, Gland; United Nations Environment Programme, Nairobi.

5.8.3 Conservation biology

Caughley, G. & Gunn, A. (1996) *Conservation Biology in Theory and Practice.* Blackwell Science, Oxford. (*A great book.*)

Hunter Jr, M.L. (1996) *Fundamentals of Conservation Biology.* Blackwell Science, Oxford. (*A well-organized and wide-ranging introduction to this subject.*)

Meffe, G.K. & Carroll, C.R. (1994) *Principles of Conservation Biology.* Sinauer Associates, Sunderland, Massachusetts. (*Addresses the major issues in conservation biology, with many helpful examples.*)

Primack, R.B. (1993) *Essentials of Conservation Biology.* Sinauer Associates, Sunderland, Massachusetts.

Primack, R.B. (1995) *A Primer of Conservation Biology.* Sinauer Associates, Sunderland, Massachusetts.

Spellerberg, I.F. (ed.) (1996) *Conservation Biology.* Longman, Harlow. (*A useful set of reviews of many of the major issues.*)

Index

Please note: page numbers in *italics* indicate tables or figures.

Action Plan, UK *104*, 105
Afrotropics 48–9
Algae *29*
altitudinal gradients 58, 59, *60*
Amazon basin 51
angiosperms *23*, 24, 25
    spatial distribution *50*, 51, 56, *57*
    value 77, 78–9
Animalia (animals) *19*, 77, 78–9
arachnids *29*
arthropods 25, *29*
Aves (birds) 33, 34, *35*, 37, 52–3

bacteria *29*
bays 62
benthic assemblages 56, 60–1
bequest values 81–2
biodiversity
    in crisis 37
    definition 1–2
    elements 2–5
    history 16–28
    importance 76–88
    maintenance 91–107
    measures 5–10
    spatial distribution 43–70
    temporal dynamics 15–38
Bioga irregularis 37
biogeographic provinces 50–1
biogeographic regions 48–9
biological control 78
biological diversity *see* biodiversity
biological species *4*
Biosphere 2 experiments 80–1
biotechnology 78, 100
birds *see* Aves
Blastozoa 8
body size 6
Burgess Shale, Canada 18, *25*
Butorides striatus sundevallis 43
butterflies 53, *54*

Cambrian *17*, 18
carbon dioxide emissions 93
Carboniferous *17*
Carson, Rachel 84, 105–6
cassava mealybug 78

cats 37
Cenozoic *17*
chordates *29*
climate change *102*
cohesion species *4*
complex systems 5–6
congruence 65–7
conservation
    ex-situ 102
    general measures 96–7
    in-situ 98–101
Conention on Biological Diversity (CBD) 1–2, 91–2, 94–105, 106
    Annex I *97*
    Article 1 (objectives) 2, 96
    Article 6 96–7
    Article 7 97–8
    Article 8 98–101
    Article 9 *102*
    Article 10 103–4
    Article 11 104–5
    definition of biodiversity 2
    elements of biodiversity 3
    Preamble 82, 83
    responses *104*, 105
    web site 14
coral-reef ecosystems 49, 56
corncrake *36*
Cretaceous *17*, 28
Crex crex *36*
crisis, biodiversity 37
crustaceans *29*

Darwin, Charles 15, 16
definition, biodiversity 1–2
depth gradients 58–61
developing countries 58, 79, 101, *102*
Devonian *17*
differences, measuring biodiversity using 5–6
dinosaurs 16
diversity–stability hypothesis 85, *86*
drugs 77–8

ecological diversity 2, 3
ecological species *4*
ecosystems
    rehabilitation 100
    responses to reduced biodiversity 85–7
ecotourism 79
elements, of biodiversity 2–5

emissions, carbon dioxide 93
endemism 51–3
energy
    consumption by humans 92–3
    supply 63, *64*
environment
    diversity and 62–4
    effects of humans 46, 63
Epidinocarsis lopezi 78
evolutionary species *4*
exotic (introduced) species 37, 100, *102*
extinction 26–8, 46
    background 28
    debt 34–5
    mass 22, 28
    random 26
    recent and future 32–7

family richness 6–8
    latitudinal gradients 53, *55*
    map of combined 67, *68*
fish
    harvesting 77, 79
    marine 56, 60–1
food 77, 80
fossil record 15–16
fragmentation, habitat 46, *102*
frost-free periods *63*
fuelwood 79
Fungi *20, 29*

generic richness *7*
    latitudinal gradients 53, *54, 57,* 58
genes 5
genetic diversity 2, 3
genus, lifespan 27
gradients in biodiversity 53–62

habitat
    disturbance 93
    fragmentation 46, *102*
    losses 46, *102*
    restoration 100
harvesting 77–8
    recreational 78–9
    sustainable 79
heron, lava 43
hierarchical organization, of biodiversity 4
history of biodiversity 16–28
    extinction 26–8
    pattern of diversification 23–6
    principal features 16–23
humans
    environmental effects 46, 63
    extinctions due to 32–3
    introductions 37
    scale of impact 92–4
    threats to existence 84
hunting 78

identification, components of biodiversity
    97–8

idiosyncratic response hypothesis 85, *86*
Indotropics 48–9
industrial materials 78
insects *21,* 25, 26, *29,* 50–1
intrinsic value 82–3
introduced (exotic) species 37, 100, *102*

Jurassic *17*
Jynx torquilla 36

land
    changes in use 93
    degradation 93–4
large marine ecosystems (LMEs) 51
latitude
    endemism and 52–3
    gradients in biodiversity 53–8
lifespan
    genus 27
    species 27, 34
lineage splitting 26
local–regional diversity relationships 46–7

mammals *22, 31,* 32–3, *34, 35*
marine ecosystems, large (LMEs) 51
marine systems
    biogeographic regions 49
    depth gradients 60–1
    latitudinal gradients 55–6
    provinces 51
    realms 47–8
marketable commodities 76–7
measures, biodiversity 5–10
medicine 77–8
megadiversity countries 50
Mesozoic *17,* 22
metazoans 17–18
molluscs *7, 29*
Monera *20*
monitoring, biodiversity 97–8
morphological diversity 8
morphological species *4*
mouse, house 37
multicellular organisms 17–18

nematodes *29*
Neotropics 48–9, *51*
numbers, measuring biodiversity using 5–6
nutrient cycling 80

option values 81–2
Ordovician *17*
organismal diversity 2, 3

Pacific Islands 33
Palaeozoic *17,* 18, 22
pelagic assemblages 56, 60–1
peninsulas 62
Permian *17,* 28
Phenacoccus manihoti 78
phyla, present-day *19–20*
phylogenetic diversity 8–9

phylogenetic species *4*
phytophagous insects 26
plants *20, 29,* 77–9
  flowering *see* angiosperms
  land *21, 23,* 24
pollinators 80
pollution, environmental 63, 65, *102*
population
  human 92
  losses and declines 35–7
Precambrian *17*
precipitation, annual *63*
productivity, primary *64,* 93
protected area systems 99–100
Protoctista *19–20*
Protozoa *29*
provinces 50–1

Quaternary *17,* 22

radiolarians *21*
rainfall, annual *63*
rats 37
realms, biological 47–8
recognition species *4*
redundant species hypothesis 85, *86*
regions, biogeographic 48–9
religion 83
restoration ecology 100
Rio de Janeiro Conference (1992) 1–2
  Convention *see* Convention on Biological
    Diversity
rivet hypothesis 85, *86*
rodents 25

Sciuridae 26
Silent Spring (Rachel Carson) 84, 105–6
Silurian *17*
snake, brown tree 37
soil formation 80
spatial patterns 43–70
  congruence 65–7
  environmental variables and 62–5
  extremes 47–53
  gradients 53–62
  scale 44–7
speciation, random 26
species 5
  concepts *4*
  fraction described 30–1
  introduced (exotic) 37, 100, *102*
  lifespan 27, 34
  losses 32–5
  number of *see* species richness
  numbers of extant 28–32
  recently/newly described 31–2, *32*

threatened 100, 101
species-area relationships 44–6
species richness 5, 9–10
  ecosystem responses to reduced 85–7
  latitudinal gradients 53, *54, 55*
  limitations/usefulness 10
  local *vs* regional 46–7
  vs generic/family richness *7,* 8
squirrels 26
subterranean diversity gradients 59–60
surrogate measures 6–10
sustainable use 79
  Convention obligations 103–4
  definition 96
  general measures 96–7

taxonomic diversity 8
temperature, environmental *63*
temporal dynamics 15–38
termites 56, *57*
terrestrial systems
  altitudinal gradients 59, *60*
  biogeographic regions 49
  depth gradients 59–60
  latitudinal gradients 53–4, 56
  realms 47–8
Tertiary *17,* 22
tetrapod vertebrates *21,* 24
tourism, ecological 79
traditional lifestyles 100–1, 104
traditional remedies 78
Triassic *17*
tropical forest 80
tropical regions 48–9, 51, 58

UK Action Plan *104,* 105
underground diversity gradients 59–60
urbanization 92

value
  direct use 76–9
  indirect use 80–1
  intrinsic 82–3
  measures of biodiversity 6
  non-use 81–3
  option and bequest 81–2
  use 76–81
viruses *29*

Wallace, Alfred Russel 91
wood 78, 79
World Conservation Monitoring Centre
    (WCMC) 14
World Resources Institute (WRI) 14
World Wide Web (WWW) 13–14
wryneck *36*